PREFACE

The following book, "Your Psychic Powers: and How
to Develop Them," was originally written for circulation
semi-privately among the members of a number of
psychical and spiritualistic Societies in New York and
vicinity, and is, for that reason, decidedly positive and
spiritistic in tone. It states facts dogmatically, and
does not attempt to defend the statements made by any
show of argument. The reader is asked to bear in mind,
throughout, the following: (1) That the present work
does not necessarily represent *my own views* in all
respects, but rather the *teachings* which are generally
accepted regarding the facts. That is to say, I have
merely endeavoured to state the traditional and accepted
theories, without in all cases endorsing these views myself. (2) For this reason, those who might be apt to
criticise certain views advanced in this book, and
brand them as rankly "credulous," "quacky" or
"spiritualistic," are hereby warned that they are de-
prived of that weapon, for the simple reason that, as
before said, the teachings put forward do not in every
case represent my own views, but the traditional and
accepted ones, which are more or less prevalent in psychical and spiritualistic circles; and I have been careful,
in nearly all such cases, to state that "it is taught,"
or "we are told," or words to that effect. I have
summarized these teachings; and the reader must use
his own judgment in selecting those which appeal to

PSYCHIC
WERS
TO DEVELOP THEM

BY

RD CARRINGTON, Ph.D.

cal Phenomena and the War," "The
nomena of Spiritualism," "Modern
ychical Phenomena," etc.

NEW YORK

MEAD AND COMPANY

1920

YOUR PSYCHIC POWERS
AND HOW TO DEVELOP THEM

CHAPTER XIV

CHAPTER XV

CHAPTER XVI

CHAPTER XVII

CHAPTER XVIII

CONTENTS

CHAPTER XXVIII

CHAPTER XXIX

CHAPTER XXX

CHAPTER XXXI

CHAPTER XXXII

CHAPTER XXXVIII

CHAPTER XXXIX

CHAPTER XL

CHAPTER XLI

YOUR PSYCHIC POWERS
AND HOW TO DEVELOP THEM

YOUR PSYCHIC POWERS

CHAPTER I

HOW TO DEVELOP

EVERY student of "psychics"—every one who has experienced "phenomena" of one kind or another, or who is more or less "mediumistic,"—desires to know how to develop his own powers and faculties so that the phenomena which come through him may be increased in power, in clearness and in excellence. It is quite possible to insure this—since we are all more or less mediumistic or psychic, and need only to cultivate our powers in order to develop them, and bring them to maturity. Development will differ, according to the character of the phenomena that we desire. Those who desire physical phenomena must develop in one way; those who desire to obtain automatic writing must develop in another; those who wish to become clairvoyant must develop in still another, and so on.

SPONTANEOUS PHENOMENA

To begin:—let me give a few general hints to those who have experienced *spontaneous* phenomena in their waking state, or who have experienced remarkable dreams, which they feel signify something—but just *what*, they do not understand. These spontaneous phenomena are often the simplest types of *mediumship*,

1

though as a matter of fact they are also an indication of
psychic power, having but little to do with true mediumistic *messages*,—that is, they are the result of remarkable
powers within ourselves. All who obtain phenomena of
this nature should make it a point, first of all, to maintain
the physical health at the highest possible standard, so
that the energies are not drained, and the body remains
healthy and the mind clear in its judgment. It is essential to reduce the amount of any stimulants which may
be taken to the lowest possible quantity, or, if possible,
omit them altogether. This applies not only to alcohol
in all its forms, but also to tea and coffee. These stimulants excite the nerves and the imagination, and often
induce "manifestations" which are not true psychic
phenomena at all, but merely the results of a disordered
nervous system. The subject should not eat too much
meat. On the other hand, fruits of all kinds, particularly acid fruits,—such as the pear, peach, plum, orange
and lemon,—are especially suitable, since the juices of
these fruits act upon the liver and tend to cleanse the
blood. Of course, these precautions are only for those
who are serious in their study, and who are determined
to obtain the best possible phenomena.

The mind should be exercised in all healthy channels.
Do not "introspect" or reflect too much on your own
inner, mental conditions. You must learn to live *outside
your head*, so to speak,—in the outer world. Do not
constantly wonder what is going on within your own
brain. If you do, you will surely lead yourself into
difficulties later on. In short, you should lead a healthy,
active life, and, between those times when you experience
phenomena, you should think about them, as applied to
yourself, as little as possible.

CONDITIONS FOR DEVELOPMENT

If you desire to obtain certain manifestations, it is not advisable to sit for them or try and obtain them for longer than twenty (20) minutes to half an hour each day. At first five or ten minutes would suffice, and this time can gradually be lengthened as you progress. This is *especially important;* and the neglect of this rule is one of the great reasons for the dangers which mediums experience later on in their development. Suppose, for example, that some one appeared to you and gave you certain advice as to your course of action. It would certainly be unwise for you to follow this advice in every case, without inquiring whether or not it would be just and sensible, and without using your own judgment when the advice was given. Even supposing that the person who appeared to you were *really* the spirit it claimed to be, there is always the possibility that this spirit may be mistaken, and the further possibility that some malicious and lying spirit was coming to you, pretending to give advice, while in reality it was only leading you astray. There is this further possibility that the figure you saw was not really a spirit at all, but merely the product of your own subconscious imagination. Often this is the case, and yet the figure has given true and sound advice! All that we are stating now is that the judgment of the individual who receives such messages, or advice, must always be exercised upon the message received. If you do not cultivate this habit, you will find that messages often become more and more insistent, when they are not followed, and will sometimes give untrue or lying information. They may even urge you to do certain things which are against your own welfare. All this can only be settled by the exercise of right judgment, and by asking the advice of those who know how much to believe in these messages. It is for this reason that the counsel and help of one who has had long training and experience in this subject is most desirable, during these early stages of medium-ship.

THE PROPER FORMATION OF THE CIRCLE

The proper formation of the circle is of the utmost importance, and upon it depends the excellence of the phenomena, and whether or not helpful personalities are drawn into your "aura" and environment. The best results may be obtained by closely obeying the following rules:—

From six to ten persons usually constitute a circle. They should range on an average from twenty to fifty years of age. Of these, half should be gentlemen and half ladies. They should sit alternately round a table, or round the room,—in case one of the party enters the cabinet. It is desirable to join hands, in order to form a "battery," so-called, and the feet should be kept flat on the floor. The circle should not last more than two hours, and not less than half an hour. An invocation or short earnest prayer should begin the proceedings, followed by slow and quiet music, which may or may not be accompanied by singing, according to the expressed wish of the "controls" or the experience of those forming the circle. The light should be subdued, but absolute darkness should not be permitted,—unless strict instructions are given to that effect. Avoid dark séances, if possible, at all times!

THE VALUE OF FLOWERS

It is advisable to have flowers in the séance room, whenever possible, as their presence is said to attract spirits in a peculiar manner. The "spirits" say that they see these flowers as "lights." Plenty of fresh air should be allowed to enter the séance room. If any member of the circle be ill, he or she should not be permitted to sit in the circle until well again.

A developing circle should meet in the same room, since this room tends to become "mediumized," or soaked with magnetic influences, given off by the sitters. The chairs on which the members of the circle sit should be wood or cane-bottomed; the use of upholstered chairs is generally inadvisable. The table round which the members of the circle sit should be free from metal. The chair on which the medium sits must be cane or wood, and, as already said, free from all cushions or upholstery.

ATMOSPHERIC CONDITIONS

Atmospheric changes play a great and important part in all mediumistic conditions. The dryer the atmosphere, the better the phenomena, as a rule. On damp, rainy days, little can be obtained. During a thunder storm, startling phenomena occasionally take place. High, rarefied air is better than that of lower levels, and for that reason a house on the side or even the top of a high mountain should be selected, if possible, in which to hold séances. Failing this, select a house which has as high an altitude as possible.

One member of the circle must, by universal consent, undertake to conduct the proceedings; to converse with the "spirits" when they appear; to arrange the sitters

in their proper places, to adjust the amount of light required, etc. His word must be followed at once, and
without question; otherwise the necessary "harmony"
will be destroyed, and the circle will fail to obtain as
good results.

Excitement in all its forms should be avoided. If
one of the circle develop mediumistic power, he should
be placed next to a more fully developed medium,—
unless instructions are received not to do so. In this
way the power is concentrated and focussed at one
point.

MISUSE OF SPIRIT ADVICE

Never attempt to use psychic-power for worldly purposes! If you do, you will invite mischievous and lying
intelligences to your circle, and the medium will possibly lose what mediumship he already possesses.

Do not sit too frequently. Every other night, at most,
should sittings be held, or even twice a week. See that
the room is not too cool and is not unduly heated. As
soon as the first manifestations have been received, encourage the "spirits" by talking to them in a natural
tone or voice, as you would if they were visibly present
in the room. Speak to them as you would if they had
returned to earth in bodily form. *Be natural*, in fact!
You will get the best results in this way.

Many of those who are interested in spiritualism are
so situated that they cannot join circles, but wish to develop alone. This is as a rule unwise, unless some one
is present who understands the phenomena which are
likely to develop and who can help and can give good
advice when required. You may do so if the following
instructions are kept carefully in mind. If you can pro-

vide yourself with a "cabinet," it would be very advisable to do so. Sit inside the cabinet on a comfortable chair and relax yourself thoroughly. Note whatever impressions come to you. Pay particular attention to your bodily feelings, no less than to your mental states. Do not exaggerate here, or let your imagination have too free play. If your legs should happen to tingle or the chair to creak, do not put these down to spiritual influence. They may be due to perfectly normal causes.

SYMPTOMS OF ONCOMING MEDIUMSHIP

For the first few evenings you will probably notice nothing much of interest, though very psychic persons begin to develop almost at once. A peculiar lightness and buzzing is sometimes experienced in the head, together with a sense of numbness in the hands and arms, and sometimes in the feet and legs. The respiration seems to become slower, and so does the heart. Tiny lights and spots and light, or dark spots, appear in the air at a distance of one or two feet in front of the subject. A peculiar pressure is sometimes experienced on the top of the head or on the base of the brain, or in the solar plexus; "swishing" sounds, as of the sea breaking upon the sea-shore, may be heard and a sensation that something inside of the head is going round and round in spirals. The head, the hands and sometimes the whole body break out into a profuse perspiration at this point.

These are the first sensations of oncoming mediumship. Very often they are not pleasant for the first few weeks, but if this period be passed, the unpleasant sensations will as a rule vanish, and the subject will then develop true mediumship of one character or another.

GETTING THE BEST RESULTS

Just here it is advisable to state that the would-be medium should not at first sit for the express purpose of cultivating any *particular* phase of mediumship. He may desire to obtain materialization, but unless he is naturally endowed in this matter, he might sit for ever and obtain nothing; whereas if he developed whatever phenomena presented themselves, he might very soon develop into a striking medium in some other line.

To return, however, to the early development of mediumship: Soon after these early impressions have been noticed, the subject may note for the first time that his mind is peculiarly susceptible to influences of all kinds. He feels as if his mind has been "skinned," so to speak, and that he is now exposed to the psychic breezes from every direction! He may become erratic and irritable and develop "moods" which he himself cannot understand. Peculiar buzzings in the head are sometimes heard, sometimes cloudlike masses seem to form in space before him, twisting and turning and moving up and down, and round about with a very irregular motion. As a rule, these clouds appear to be of the consistency of vapour, though they may in time become more and more solid, until they become built up into definite forms. Of this however later.

EARLY SIGNS AND EXPERIENCES

At this phase of the development the subject may feel cool breezes blowing upon his hands and face from various directions—breezes which appear to be perfectly physical in character. He may also experience a peculiar sticky sensation on his hands and face as though cobwebs were applied over the bare skin which is exposed. This cobwebby sensation is very common, and is not limited, as many think, to mediums who obtain materialization.

COLOURS AND VOICES

In the early stage of the development, mediums very often see colours of various shades and hues in space before them. They are unable to tell whether or not these colours have any definite shape or outline; they seem to possess an odd, irregular shape of their own; something like a large blot of ink. At this stage also many psychics see faces of friends and relatives, either living or dead, just as they are falling asleep or as they are awakening in the morning—more usually the former. They also see many strange faces. These may be mere vague images or clearly outlined. Instead of the faces they may hear voices, speaking—and the first thing which these voices generally say is the name of the subject himself. After this the voices may become more and more clear and intense, but such phenomena should be permitted only at stated times, because if they are allowed to develop whenever they may be experienced, trouble may result. Many odd and grotesque figures and shapes may present themselves to the mind's eye at this stage of development. These shapes may be highly coloured or may be almost colourless, seeming to be made of the air itself, yet somehow separated from this in outline. Many of these images are symbolic, though as a rule a few of them are recognizable. More often they represent curious patterns and figures, such as roses, circles, outlines of patterns such as may be seen on the wallpaper and occasionally weird and horrible images flash into the mind, to be gone again the next instant.

UNPLEASANT EXPERIENCES

If these manifestations develop an unpleasant character at this time they should be checked instantly. The
subject may do this in several ways. First of all he
should build up his physical health. Second, he should
see to it that he obtains plenty of sleep. Third, he
should exercise his brain as little as possible on anything
of this unpleasant character. Fourth, he should keep
busily occupied in material, practical things and leave
himself no time to ponder and dwell upon these unpleasant occurrences. Fifth, he should avoid by all means
day-dreaming and never allow the mind to become passive
or absent-minded. He should cultivate his objective attention and interest, in short, and focus his whole personality, as it were, between his eyes, so as to have it under
thorough control. If he does this, and refuses to sit for
development for a short time, he will find that these
early unpleasant symptoms (should they develop) will
soon wear off; and this advice holds good at any stage
of the development of mediumship.

EXAGGERATION AND IMAGINATION

Many of those who develop psychic phnenomena are
inclined to exaggerate the importance of the manifestations they receive during the early stages of their
mediumship. Everything seems so new and strange to
them, so remarkable, so unaccountable, so beyond the
experience of the average person, that they feel bound to
tell it to every one they meet and usually it loses nothing
in the telling! They fail to realize that every medium
who has been developed has gone through these same

early stages, but has progressed beyond them years before! In observing these phenomena in yourself you must be very careful to distinguish between the facts which really occur, and the phantasies of your own excited imagination, which is inclined to extend and amplify these facts beyond all recognition. Thus, suppose a blurred outline of a face presents itself to you; the next day you meet your cousin on the street. You instantly come to the conclusion that the face you saw was that of your cousin,—while, as a matter of fact, it might not have borne the least resemblance to him. This is a very simple case, but will serve to explain the point in question.

WHY AND HOW YOUR POWER MAY BE LOST

If you obtain such phenomena, you must be very careful not to exaggerate them, for if you do you will quite possibly lose the real sensitiveness that you are beginning to acquire, and this will be replaced by the products of an overexcited imagination.

This is a truth, well known, as you may see by the following quotation from a work which appeared in 1813, entitled "Animal Magnetism" by Deleuze, for in it he says: "Do not press the somnambulist too much, for if you do you will gain nothing; you will even lose the advantages which you might derive from his lucidity. It is possible that you could make him speak upon all the subjects of your personal curiosity, but in that case you will make him leave his own sphere and introduce him in yours, he will no longer have any other resources than yourself, he will utter to you very eloquent discourses, but they will no more be dictated by the external inspirations, they will be the product of his recollections or of his imagination. Perhaps you will also rouse his vanity and then all is lost, he will not re-enter the circle from which he has wandered. The two states cannot be confounded.''

The student, who cultivates mediumship should, therefore, be careful to preserve a clear head and a modest estimate of his own phenomena. If he does, he will doubtless progress rapidly and favorably.

CHAPTER II

"HARMONIOUS CONDITIONS"

If we exert ourselves in any way whatever, we desire certain "conditions," in order to bring our powers and faculties into play to the best advantage. If we are undertaking to perform any feat of physical strength, of intellectual or spiritual achievement, we desire to be free from care and worry, distraction and irritation— to be enabled to centre and focus the whole of our energy in the channel desired. It is the same with mediumship.

CONDITIONS FOR THE EXERCISE OF PSYCHIC POWERS

Professor Flourney, of Geneva, writes in this connection: "As to the influence of various physical and mental conditions upon the exercise of mediumship, my correspondents are unanimous in condemning as absolute hindrances or at least grave obstacles to the production of phenomena, all such causes as physical exhaustion, disturbing emotions, uneasiness, absorbing thoughts, fatigue, enervation, etc. The conditions required for the successful exercise of mediumistic powers are the same as for the voluntary exercise of any other power, —a state of good health, nervous equilibrium, calm, the absence of care, good humour, sympathetic surroundings, etc. Many insist upon moral elevation, purity of conduct, noble aspirations, altruism, etc.,—saying that these things strengthen mediumship, while the lower sentiments such as cupidity, pride, jealousy, etc. are the

13

cause of much loss of power. Others have insisted that certain physical conditions have a propitious effect, —silence, semi-obscurity, good ventilation, fasting, etc.''

NECESSITY FOR CONDITIONS

Those who do not understand the laws of Spiritualism have contended that the ''conditions'' demanded by mediums are often absurd, for the reason that they permit trickery. If the conditions permit the practice of fraud, they should not be allowed. Beyond this, any conditions required by the medium should be granted, for the medium alone is the one to know what these conditions should be. Mediumship, doubtless, has its ''conditions''—its own psychic laws—just as any other exercise of the inner powers. Many sceptics do not see this. They say: ''If you can produce these phenomena, you must be able to produce them at any time, just as we can always produce the same effects in a chemical or physical laboratory! Why all this fuss about conditions, etc.?'' But they fail to take into consideration *human nature*, and the fact that psychic laws and physical laws are different. We can easily prove this.

''CONDITIONS'' IN ART

Take any musical composer or any artist who paints, and seat him at a table with instructions to compose a sonata or paint a wonderful picture, within half-an-hour! Suppose that during all the time the work is in progress, noise and flurry is constantly going on in the same room, the desk at which the artist works is being shaken, children are continually running in and out of the room, etc. Do you think that, under such conditions, a masterpiece in either music or art could be produced? Could a poet compose a sonnet under such conditions? Certainly he could not! The exercise of mediumistic power is assuredly as delicate, as subtle, as refined, as easily disturbed as any of these productions of the genius of man. How absurd, therefore, to pretend or contend that mediums should be able to exercise their powers, whenever they want them, under any conditions! And to contend, further, that if they fail to do so they are therefore frauds and humbugs!

For the successful exercise of mediumship or psychic power in any direction, the essentials which have been mentioned above must be fulfilled, as well as any others which the medium may feel are required. These must by all means be granted, for if they are not, it is highly probable that no phenomena at all will be obtained.

HARMONY ALL-IMPORTANT

Harmony is the keynote of successful mediumship,— harmony of physical, mental and spiritual life. This is only carrying to its logical conclusion what we observe every day all around us. Have we not all felt, immediately upon meeting certain persons, that they were attractive or repellant to us? We felt either drawn or repulsed inwardly for no reason that we could define. Many theories have been advanced to explain this fact, but the most probable is that, surrounding each individual, there is an "aura" or psychic atmosphere which surrounds him like a halo or sheath, extending some distance outward from the body and varying with the individual temperament, emotions and the physical and mental health. If these auras are sympathetic, if they blend one with another, then we have attraction, leading

in many cases to "love at first sight;" if the opposite
conditions exist, we have instinctive dislikes which are
generally correct. As the poet said:

> "I do not like you Dr. Fell,
> The reason why, I cannot tell."

MENTAL HARMONY

Next to physical harmony comes mental harmony, and
here is a wide field for observation and experiment.
All Spiritualists know that persons of certain temperaments must be excluded from serious circles, if the
better class of phenomena are to be obtained. Such
persons include the flippant, the arrogant, the unduly
sceptical, the frivolous, etc. In addition to this, however, finer and more subtle points of mental harmony
must be adjusted in our mental scales. It is advisable,
whenever possible, to bring together persons having more
or less the same point of view, interests and sympathies.
Sympathetic people always obtain better phenomena than
the extremely intellectual ones. In the latter, the mind
is, so to speak, hard, unyielding, and tends to build up
a wall or barrier between itself and the medium, which
it is difficult or even impossible for the latter to break
through. We have known of several cases in which
mediums were unable to obtain any results at all for
individuals of the very intellectual and, so to speak,
critical type of mind, whereas they could obtain an
abundance of striking manifestations for the sympathetic
and more congenial natures.

EXCESSIVE GRAVITY

At the same time, extreme gravity and seriousness
on all occasions is to be avoided. Every person who

investigates spiritualism should see to it that he pre-
serves throughout his sense of humour, and his continued
contact with, and interest in, the things of this world.
If he does not do this, he is liable to become unduly
swayed and overbalanced by the messages which are
given to him, and by the startling and at first sight
almost appalling fact that communication with the spirits
of the departed has really been established! If he does
not preserve his balance and common sense at such times,
he is liable to become not only unduly credulous but even
to "fly off the handle" altogether, and his mind may
in some instances actually become unhinged. Be
careful, therefore, to keep the "Compartments" of your
mind watertight, as it were, and not allow your interest
for the "things spiritual" to overflow and swamp your
interest in things material.

SPIRITUAL HARMONY

Next we come to spiritual sympathy and harmony,
which is perhaps most important of all in the formation
of successful circles. This would include an interest
in spiritual things, aspirations, benevolence, sympathy,
a more or less religious turn of mind, tolerance, and the
ability to see things from the standpoint of another,
this being sympathy in its broadest sense. It is the
blending together of a number of temperaments of this
character which constitutes the successful circle. The
reason for this harmony and delicate adjustment of conditions may be seen readily enough by a reference to the
phenomena not only of the mental but of the physical
world. For instance, if you set into vibration a tuning
fork, this will emit a note of a certain pitch, another
tuning fork, distant many feet from the first; will in-

stantly vibrate in unison if the two are attuned one to the other. But unless the tuning forks are adjusted at *precisely* the right pitch, they will not respond, and a thousand tuning forks may be placed around the room but none of them will respond in any way to the vibration of the first. This crude analogy, drawn from the physical world, will show us how essential *harmony* is; and, if this be true, in material phenomena, more certainly is it true in the mental and spiritual realms.

HARMONY AND VIBRATION

Every individual is said to vibrate at a certain rate; this is his own "pitch," so to say, and no two human beings are alike. This definite rate of vibration doubtless corresponds to the personality of the individual, and, though no two can be absolutely alike, those who *approximate* each other the nearest, would be the most sympathetic and would be the most drawn one to the other. And if this is true of spirits incarnate here in this life, it is doubtless true when applied to the relations between our own spirits and the spirits of those who have passed over. There is an old saying that "Like attracts Like." If the tone of the circle, and of the individuals composing it, is high, the aspirations elevated and pure, that circle will attract to it "spirits" from the other side, having the same vibrations as itself. The circle will, in fact, only come in contact with good and not evil spirits. Certainly, there are exceptions to every rule; but the above is the *general* law which may be stated in broad terms as true. For, were this not the case, we might contend that no such thing as justice existed in the Universe, and that Chance and not Moral Law held all in its sway. But we know

that this is not the case, inasmuch as we feel assured in
our heart of hearts that beauty, truth, and justice are the
foundation stones upon which this universe is built.
"We might rightly suppose that this is in every case the
truth, and that a circle formed by serious-minded investigators, having in view only the highest and best
motives, would draw to them helpful and loving spirits
from the great beyond. And the history of Spiritualism
proves this to be the fact."

HOW TO FIT YOURSELF FOR A CIRCLE

The method you should follow to fit yourself most
effectually for becoming a member of one of these advanced circles is as follows: You should perfect and
make as wholesome as possible the physical body in
which you live; this is brought about by paying particular attention to the diet, and by taking an abundance
of exercise, deep breathing and frequent baths. Many
spiritualists have become vegetarians with this object in
view, and also non-smokers and abstainers from alcohol.
Tea and coffee are also debarred in some quarters; but
such strict measures are usually advised only for those
who are striving for individual spiritual perfection,
and need not necessarily be followed by one who is a
member of a large circle. Of course such measures cannot fail to benefit an individual in any case.

THE KEYSTONE OF THE ARCH

Cultivate cheerfulness, altruism, and a simple, wholesome point-of-view,—banishing fear as you would the
Devil, and never allowing it for a moment to dominate
or enter into you! Preserve a sane religious balance,

and endeavour in every way possible to cultivate sympathy for the point-of-view of others, no matter how prejudiced and narrow it may be. Keep your mind *lifted up*, elevated; and, as Andrew Jackson Davis said, "Under all circumstances keep an even mind." If you do not naturally possess it, cultivate an insight into things spiritual, and above all true benevolence and sympathy. This is the keystone of the arch erected to, and supporting, self-perfection.

CHAPTER III

FEAR AND HOW TO BANISH IT

WE are our own greatest enemies. We create the majority of the ills from which we suffer! In psychic investigation, more people have suffered from fear than from any other depressing emotion; but, in nine cases out of ten, these fears have been perfectly groundless, and the subject has had all his fears and worry for nothing! He has crossed his bridges before coming to them. Were he to reflect for a moment, he would find that the terrible things he feared very rarely came to him; that the majority of the experiences which he actually went through were of such a nature that he needn't have feared them at all.

FEAR WRECKS, FAITH SAVES

Fear is not only useless (for the reason that it prevents nothing) but it is actually harmful from this double standpoint: In the first place, it helps to induce the condition we are fearing. As Job said, "That which I greatly feared has come upon me." He thought about and dreaded certain conditions so much that he doubtless created them, while, had he not done so, they would never have come upon him. Professor William James gives us a very good illustration of the way in which fear sometimes brings about its own fulfilment. He says: "Suppose that, for example, I am climbing in the Alps and have the ill luck to work myself into a position from which the only escape is by a terrible

21

leap. Being without similar experience, I have no evidence of my ability to perform it successfully, but hope and confidence in myself make me sure that I shall not miss my aim, and nerve my feet to execute what without those subjective emotions would perhaps have been impossible. But suppose that on the contrary the emotions of fear and mistrust predominate, or, suppose that I feel that it would be sinful to act upon an assumption unverified by previous experience, why then, I shall hesitate so long that at last, exhausted and trembling, and launching myself in a moment of despair, I miss my foothold and roll into the abyss. In this case (and this is one of an immense class) the part of wisdom clearly is to believe what one desires, for the belief is one of the indispensable preliminary conditions for the realization of its object. There are then cases where faith creates its own verification. 'Believe and you shall be right, for you shall save yourself.' 'Doubt, and you shall again be right, for you shall perish.' The only difference is, that to believe is greatly to your advantage.''

The obvious lesson to be drawn from this, is, that you should not fear the unknown or unseen until you have had just cause to do so. If you do, it will predispose you to experience the very manifestations you most dread.

EVIL EFFECTS OF FEAR UPON THE BODY

In the second place, Fear has a destructive and depressing effect upon the body. It depletes the vitality, lowers the respiration and doubly incapacitates you from performing any serious, rational work or carrying on any rational common-sense train of thought. Fear, therefore, is certainly to be avoided, for "it helps nobody and harms everybody." But, the reader, may object, "I cannot control my fear so easily, it is a thing beyond my power, I do not pursue fear, it pursues and overtakes me." To a certain extent, this may be true.

There are two kinds of Fear, the unreasoning instinctive fear, and the conscious, reflective fear. The former is a relic of our lowly ancestry, and is shared by all the higher animals. We cannot help that, but such fear, as a rule, is only momentary and is over in a few instants,—we have the impulse to flee, etc., which demands immediate expression, but this instinctive fear may be overcome by the mind. Our reason tells us, upon second thought, that we have no cause to fear, and we stop abashed and ashamed of ourselves. This is not the fear which we have to combat, as a rule, since it is bodily rather than mental, and of short duration.

THE KIND OF FEAR TO FEAR

The conscious, mental fear is that which bothers us and which we should learn to cure. We are sufficiently advanced in civilization and in the understanding of things spiritual to know that all is natural. Nothing is supernatural. Even if a spirit returns to us, that is a natural event, though it may not be a common or ordinary event, and for this reason, we call it "supernormal." But why should we be afraid of the spirit of a dearly beloved friend or relative, or even the spirit of a stranger coming to us in this way, any more than we should be afraid of it when coming to us in the flesh? It is the same spirit,—in one case possessing a physical body, in the other case animating only an ethereal body. Of what is there to be afraid? Spirits

are but human beings, such as ourselves. We are spirits
here and now, just as much as we ever will be. Spirits
are, in fact, human beings who have passed through a
certain experience, called "death" and, as Professor
Minot Savage said, "They are just folks." Why, therefore should we be afraid of them?

THE POWERS OF DARKNESS

We must school the mind to reflection and, by due
exercise of the reason and the will, not to be afraid of
such happenings, but rather to accept them and be
thankful for them, and to treat them either as scientific
happenings or as spiritual events of great significance
and help. In either case, there is truly no cause to fear.
It is true that in the case of many persons, darkness
brings with it a peculiar sense of dread, which is experienced by nearly all children and which is to a certain
extent shared by many animals. A dog will go to the
door of a dark room, peer in and slink away. Even
insects often refuse to go into dark places. The cat
alone seems to enjoy the uncanny sensation which accom-
panies darkness, and we know that cats are proverbial
"ghost lovers," while dogs are the reverse!

It may be that there is more truth in this belief than
many realize. We know that the orthodox devil was
known as "The King of the Powers of Darkness," and
all evil things are associated with that state. On the
other hand, Jesus was said to be "The Light of the
World" and *light* always accompanied spiritual manifestations—as it does today. The expression made use
of by Mr. Hamlin Garland some years ago in his book
"The Tyranny of the Dark" may, therefore, have a
certain foundation. There are perhaps "principalities

and powers'' which can operate more freely and fully
in the dark than in the light, but only if they are
allowed to do so by the fear and the attitude and mind
of the person experiencing them. We remember that,
in "Pilgrim's Progress," the travellers were repeatedly
warned that no harm could come to them so long as
they faced their spiritual enemies. And we must remember the words of the greatest of all psychics, "Resist
the devil and he will flee from thee.''

All we have to do, therefore, in order to prevent the
domination of any evil thought or power is to fight it,
resist it, meet it strongly and courageously with calm-
ness and decision and it will melt before your attack like
dew before the rays of the morning sun.

EVIL EMOTIONS

Mr. Horace Fletcher, in his little book on "Happiness"
says some very good things regarding fear, which he
defines as "an expression of fear-thought.'' "Fear-
thought,'' according to this author, is "The self-imposed
or self-permitted suggestion of inferiority. It is both
a cause and an effect of selfishness. It is the tap-root
of evil.''

The body is a mirror in which all states of the soul
are reflected. Perhaps the most extensive of all the
morbid mental conditions which reflect themselves so
disastrously on the human system, is the state of fear.

Dr. Hack Tuke, in his book, "The Influence of the
Mind Upon the Body,'' cites a number of well-authenticated instances of disease having been produced by
fear or fright. Insanity, idiocy, paralysis of various
muscles and organs, profuse perspiration, turning the
hair grey in a short time, baldness, nervous shock fol-

lowed by fatal anæmia, mal-formation of the embryo, and even skin and other diseases, apparently more removed than these from the effects of the mind, were traced to the effects of fear and other mental disturbances. He pointed out, also, that epidemics, such as cholera, small-pox, diphtheria and other malignant diseases, obtain a footing in a community largely through the fear of the inhabitants, and that hundreds and even thousands of persons fall victims to their own mental conditions.

HOW FEAR CAUSES SICKNESS

How does fear operate upon the body to produce sickness? Largely by paralyzing the nerve centres, especially those of the vaso-motor nerves—thus producing not only muscular relaxation, but capillary congestions of all kinds. It is an interesting fact that fear and all depressing emotions of a similar nature serve to constrict or contract the body—while mirth, love, altruism, and all the higher emotions serve to produce both physical and mental relaxation—opening up the mental and physiological doorways of the organism. The term "frightened to death" is not a mere expression, but is founded upon valid physiological and psychological laws.

A Southern physician has reported an interesting case. It was that of a big, burly negro, who supposed that he had been fatally shot. Fear had seized him with tremendous power, he shook like an aspen leaf, he bordered on the state of collapse and death seemed imminent. Not finding any blood, the examining physician ordered all his clothes removed and, while he was being undressed, a flattened bullet fell upon the floor. The doctor exhibited the bullet to the frightened patient, explaining that he had had a miraculous escape, whereupon his circulation was immediately restored, his countenance improved, his temperature became normal and the look of life returned to his eyes which had been fixed with the gaze of death, while a broad grin crept over his face. The negro got down from the operating table and dressed, apologized for the fuss he had caused and walked home!

FEAR IS CONTAGIOUS

Fear has the peculiar power of being extremely contagious. Under the proper conditions, fear manifested by one person is instantly communicated to the entire company. They feel little chills run up and down their spines, their hair begins to stand on end, and a cold perspiration breaks out here and there over the body. This shows the profound effect which this emotion has upon the bodily functions, and also how easily it may be acquired without reason. Fear has the power of almost stopping the heart and paralyzing the entire nervous system. A peculiar fatigue is also caused by fear, as has been proved by delicate experiments. A natural and normal way to overcome fear under such conditions is to open the mind to natural faith and normal trust. Let the psychic forces be allied with faith and health, let fear be finally and for ever cast down and banished from the mental domain. This may often be brought about by reasoning, though an effort of will is generally necessary also. A determined opposition accompanied by trust, faith in wise protection, faith in your own powers and in the help of friendly spiritual monitors, are of the greatest use and benefit in overcoming this great monster "Fear."

THE FEAR OF EVIL SPIRITS

Many people are afraid of "evil spirits," being alarmed lest they should influence them against their will and cause them to do certain things which they would not normally care to do, even to the point of obsessing them. There is a real danger here, to a certain extent, which will be dwelt upon and explained in the chapter on "Obsession." But let it be pointed out that the only way to prevent such things, is to keep up a normal, healthful resistant attitude of mind and not to give way to fear, which would be doing the very thing to invite attack. Let us recall once more the words of Job in this connection. So long as the sea-walls or dykes of Holland are sound and unimpaired, the ocean is kept within its proper limits and cannot break through and flood the land as it sometimes does when these walls are destroyed. As we know, a tiny little hole through which the merest trickle of water can pass, will, unless repaired, soon become a wide crevice and then a roaring torrent. The most important thing to do is to check this in its inception, for it is easy to prevent the ingress of the water if taken in time; better still, it would be far easier to keep the sea-walls in such repair that accidents of this kind would be impossible, for "prevention is better than cure."

Applying this to the case before us, we can see that the very first symptoms of fear must be checked as soon as they arise, for, if they are allowed to continue, they will spread and work havoc in the mind, just as the waters would work havoc upon the land. The thing to do is to keep the mind so guarded, strengthened and repaired by healthful exercise, intelligent cultivation and

control and the exercise of the will, that fear can never
batter down its ramparts, and, even should it attack
the citadel of the mind, it would be quite unable to find
a lodgment within this impregnable fortress.

THE FEAR OF BEING HYPNOTIZED

What has been said applies also to the action of hypnotic influences, which many persons fear greatly.   They
are afraid of being hypnotized by some "distant operator," and this fear sometimes becomes with them a veritable phobia, so that we occasionally find insane asylum
patients who have become completely unhinged on account of this fear.   We can see from this how useless,
how exceedingly harmful fear of this character is, and
it is more than useless, it is ridiculous!   No one can
be hypnotized against his own will by a distant operator
in this way, as many suppose.   If they feel influences
of this character, these feelings are the result of their
own disordered imagination, and are not due to any
outside influence whatever.   An individual really hypno-
tizes himself, the operator directing his own mental
powers into certain channels so that this is brought about.
If he resists the suggestion, as every one can do at first,
it is impossible for any one to hypnotize him.

WHY SUCH FEARS ARE GROUNDLESS

The only way in which a person can be hypnotized
from a distance is the following: If an operator has
hypnotized his subject a great many times and repeatedly
suggested to him, when in the hypnotic trance, that he
is becoming more suggestible, that he can easily go off
to sleep, that he has only to think of the operator in

order to fall asleep, etc., he may succeed in making the subject so sensitive, after a certain length of time, that this condition is really brought about. The subject tends to fall into trance on the slightest provocation. But such cases are abnormal and are rarely met with, and, as I have just said, this condition cannot be brought about until the subject has been hypnotized several times and these suggestions given to him. These are well known facts which any experienced hypnotist will sustain. This being so, it may readily be seen how absurd it is to fear telepathic suggestion from a distant operator, whom perhaps you have never seen! It is entirely illusory, and you need in reality have no fear whatever in this connection. The will if exercised is supreme!

CHAPTER IV

THE SUBCONSCIOUS

IN this chapter I shall take up and try to make plain to the student the nature and functions of the subconscious mind. This is the greatest of all stumbling blocks to many spiritualists. Its possibilities, and at the same time its limitations, should be made clear to the student at the beginning of his studies, otherwise he is sure to get in trouble later on, not only with himself and with the phenomena he is studying, but with all persons who discuss these subjects with him, and try to persuade him that the whole of Spiritualism may be accounted for by the powers of the subconscious.

WHAT IS THE SUBCONSCIOUS MIND?

First of all, what *is* the subconscious mind? We do not know exactly, but a great deal has been found out concerning it within the past quarter of a century. Twenty years ago, when Thomson J. Hudson wrote his famous work, "The Law of Psychic Phenomena," very little was known of the subconscious. Nearly everything which has been discovered about it has been learned since he wrote. His attitude is doubtless well known to the majority of my readers. It is that man has "two minds," the conscious and the subconscious; or, as he preferred to express it, the "objective" and the "subjective" minds. The first of these is the conscious mind,—the every day, reasoning mind; the second is that vast realm in which occur the phenomena of

31

dreams, hypnotism, insanity, hysteria, clairvoyance, telepathy and all kindred psychic phenomena. He placed the "objective" mind in the cerebrum, or fore part of the brain, and the "subjective" mind in the cerebellum, or hinder part of the brain.

TWO MINDS, OR ONE?

But this dual conception of the mind is today given up by practically all psychologists. They admit that the mind is, in a certain sense, dual, but it is believed that both minds are in reality *one*, a part of which is conscious, and of the greater part of which we know nothing. The analogy of the iceberg has often been used. A small percentage of this emerges above the water, and this we see and know; but the greater part of the mountain of ice is below the surface, and this we do not know through our senses. Yet it is all one iceberg! In the same way there is only *one mind*, but when the searchlight of consciousness is turned upon certain areas, those areas become illuminated, and we "know," or are conscious of, those parts. All else remains in the dim obscurity beyond, in the great storehouse of the subconscious mind.

THE POWERS OF THE SUBCONSCIOUS

It may easily be proved that the subconscious mind acquires far more information, even through the senses, than does the conscious mind. The following simple experiment will prove this: Lead a person into a strange room and ask him to observe as many things in it as he possibly can. Suppose he remains five seconds in that room. He is then quickly removed and the door shut. If, now, he is asked to tell you all the things that he remembers having seen, he will probably be enabled to remember ten or fifteen of them; but if you were to hypnotize that person, he would then describe to you, under hypnotic influence, forty or fifty things which were in the room. This shows us that the subconscious mind, which we reached through hypnotism, has been able to perceive or "take in" many more things than the conscious mind. This happens to us every day.

DREAMS

It is the same with *dreams*. If, when out walking, we should happen to drop a brooch-pin or a piece of money, we might be totally unconscious of the fact, but the subconscious mind would perceive and record it. That night, in sleep, we might have a dream in which a figure appeared to us and told us that the article had been lost, and that it would be found in such-and-such a place. On looking, the next morning, sure enough there it was!

Here we see that the subconscious has perceived a certain fact which the conscious mind did not notice. For long it was thought that this power signified some supernormal faculty of the subconscious mind, but in most cases it is not necessary to suppose this, for, as the last experiment showed us, the subconscious mind takes in many things which the conscious mind does not, and only the most striking and interesting facts rise into consciousness. Of the thousands of events going on all around us every day, we perceive but a few; all the rest are ignored, though they are lodged within the great mental store-house within us.

THE MEMORY OF THE SUBCONSCIOUS

The powers of the subconscious mind are indeed great. It forgets nothing, and facts which have entirely slipped from the conscious mind are retained within it, and may be recalled years later, or may suddenly flash into the memory of their own accord. They may come into the mind in the form of some simple thought or memory, just as any other thought or memory would; or, they may come to us in more startling form. They may be, as we say "externalized;" that is, projected upward from the subconscious into the conscious mind, forcibly and dramatically, as a bomb-shell might be exploded within it. In these cases, the thought may strike us as coming wholly from without, and not from within ourselves at all. One or two examples will make this clear.

You have mislaid a book; you cannot remember where it is. The natural process would be to "recollect." In the case of an individual who is psychic or mediumistic, the "externalization" may take more startling form. He may hear a voice telling him to look under certain papers upon the library table, and, sure enough, upon looking there, the book is found! Or, he may have a mental picture of himself leaving the book in that place. Or, he may feel a hand, gently pushing him in the direction of the table. Or he may see a figure standing before him and pointing to the hidden book. In all these cases it is improbable that the voice, the touch, and the figure, were *real*,—that is, that they came from some spirit friend. They *may* have done so, but it is true that, in many cases at least, they are methods by which the subconscious mind "externalizes" or reproduces its hidden memories in dramatic form, just as they are reproduced in dreams or in visions of the crystal ball.

THE PSYCHIC DIAPHRAGM

The subconscious mind, therefore, may be looked upon as composed of a number of *strata*, like a layer-cake, which are, normally, more or less separated from one another by a sort of psychic membrane or "diaphragm" which is impervious. At times this "psychic diaphragm" becomes thinned. In that case, we remember our dreams of the previous night, or we have wonderful constructions of genius, the productions of musical prodigies, etc. The subconscious mind works out the problems, and the finished product is projected into the conscious mind in its completed form. That is why it appears to us so marvellous. On the other hand, if a part of the subconscious mind is diseased, as may sometimes happen, then we have hysteria, obsession and insanity.

It will be seen, therefore, that both good and evil may result from this thinning or puncturing of the "psychic diaphragm," separating the conscious from the subconscious mind. If the mind be healthy, and is kept so, only good will result. Psychic powers will be cultivated and helpful advice will be given to the subject thenceforward. If, on the other hand, the mind becomes in any way deranged or diseased, then harm may result, and the individual may be sorry that he has ruptured this dividing diaphragm instead of preserving it intact. It is all a question of care, good health, good judgment, and a healthy psychic, mental and physical life.

Once this psychic membrane has been, so to say, "punctured," it is very difficult to heal it up again, and great care must be exercised in developing these subconscious phenomena. We shall discuss this more fully, however, in the chapter devoted to "Obsession."

The subconscious mind should be made our friend and not our enemy. We should train it carefully, for, though it is a good servant, it is a bad master! It should always be kept in check and dominated and controlled by the conscious mind. When this is the case, all goes well.

HOW THE SUBCONSCIOUS RECKONS TIME

The subconscious has, among other faculties, the power of reckoning time in a most remarkable manner. Many of my readers have doubtlesss conducted the following experiment for themselves. On going to bed, you have said to yourself, "Now I wish to awake tomorrow morning at seven o'clock promptly, because I have such-and-such a train to catch." There is no alarm clock in the house, but, promptly at seven, you awake! That this is no mere chance coincidence has been proven by a number of cases, which have been collected, and the fact has also been proven experimentally on hypnotic subjects. Thus, they have been told that in (say) 9,750 seconds they would perform a certain action: then they were immediately awakened. As soon as awake, they knew nothing of the suggestion which had been given to them, and nothing of the action they were to perform, and yet precisely in 9,750 seconds they performed the action in question! We see, therefore, that the subconscious mind has the faculty of reckoning time in a very remarkable manner, and this is but one of its mysterious powers.

THE SUBCONSCIOUS RULES THE BODY

Another of its remarkable manifestations is the power which it possesses over the bodily organization. By means of suggestion, the pulse has been raised or slowed, the temperature has been elevated or lowered, the various secretions of the body have been altered, and many similar phenomena which are well-known to any one who has read upon this subject.

One of the most striking cases, doubtless, is that of Mlle. Ilma X. A pair of cold scissors was applied to her chest, and it was suggested that these were red hot and that they were burning the flesh. In a few moments an angry red mark appeared, corresponding to the shape of the scissors, and the next day a genuine blister had been created which took several days to heal! Here we see the power of the subconscious mind in affecting the body, and even the local tissues to a remarkable extent.

If this is true, and the body can be harmed in this way, it can doubtless also be cured. We here enter the field of suggestion and psychotherapy, which will be treated more fully in Chapter XXX.

HOW TO GIVE YOUR OWN SUBCONSCIOUS MIND SUGGESTIONS

One of the best methods of treating yourself is by suggesting certain desirable things just as you are falling off to sleep. Thus, if anything is wrong with you physically, mentally, or spiritually, suggest to yourself, the last thing at night, as you are falling to sleep, that all will be well, that the trouble will be removed during

the night, that you will wake up refreshed and invigorated, that there be no pain, no unpleasant feelings or
emotions in the morning, etc. Suggest, in fact, whatever
you desire to have accomplished, and you will find that,
during the night, this will have been effected and that
your bodily or mental ills will have disappeared as the
result of your auto-suggestion during the hours of sleep.
Here again we shall be enabled to see the remarkable
powers of the subconscious mind brought into play and
clearly demonstrated.

SPIRIT MESSAGES AND THE SUBCONSCIOUS

Now these faculties of the subconscious mind explain a
certain number of "spirit messages," which are received
at séances. Let me illustrate this in the following manner: Just before leaving your home to join a circle,
you glance at the evening paper. Your attention has
been attracted to the leading articles; apparently you
have seen nothing else. At the séance, that evening,
the name of a friend of yours is spelled out, and the
announcement that this friend has been killed that day,
by falling from the fourth story of his residence! At
first sight this seems a very good "test message," but
on going home and again looking at your evening paper,
you find a small article tucked away in the corner of
the paper, stating these facts. Therefore, probably what
happened was this: Your subconscious mind perceived
and "took in" these facts without their even rising
to consciousness, and, at the séance, they were given
out either by yourself, or by the medium who obtained
them from your mind by telepathy. In this way many
messages have been explained and shown to be due to
the workings of the subconscious mind, and not to spirits

at all. We must, therefore, always be on our guard
against these possibilities.

THE STRUCTURE OF THE MIND

The older conception of the human mind was that it
was a single entity, an individual thing, a sort of
"sphere" incapable of division. This was, in fact, one
of Plato's main arguments for the Immortality of the
Soul. Unfortunately modern science has destroyed this
illusion. We now know that the human mind is a com-
posite and not a simple thing. To use a rough analogy,
it has been proved that the mind is somewhat like a
rope, composed of a number of strands, twisted together.
Under normal, healthy conditions, this rope remains
one; the strands are united; but under certain abnormal
states or conditions, these "strands" may be divided up
into several groups, and they would all pull in different
directions. What holds these strands together normally?
First, good health; then cheerfulness, attention, concentration, will, and an interest in objective things.
What favours this disintegration process,—this "disso-
ciation of the mind," as it is called? The exact opposite of all this—a run-down or fatigued condition, introspection, and particularly all continued subjective
practices and the too-passive attitude of the mind. If
we lose contact with, and interest in, the objective world,
if we go "inside our heads" and spin romances and
dream day-dreams to too great an extent, if we gaze
blankly into space, thinking of nothing in particular,
if we allow the mind to become too passive and do not
exercise our intellect in a normal, healthy manner, this
disintegration is likely to take place. The "strands"
of the rope become separated, and then the mind may

go to pieces and "spirit obsession" and even insanity may result.

HOW TO HEAL A SICK MIND

Of course this is only a crude analogy. The mind is *not* like a rope and cannot be divided into strands in the same way, but it is an analogy which will help us. The only way to heal and restore a mind in this condition is to weave or weld together these separate strands and bind them up again into one solid single "rope," as it were. This may often be done by hypnotism, but great care must be exercised in doing this, for if it is not rightly applied by an expert operator, the mind may become still more disintegrated, and "the last state of that man shall be worse than the first."

Now, these separate strands of the mind (to return to our analogy of the rope) may form different "selves." Each self may possess a certain identity and individuality of its own, and they may all pull in different directions,—that is, they may all exercise their own functions and powers, and think their own thoughts. There is no *one* self any more; it has "gone to pieces." These various selves may alternate one with another in the same individual, and then we have those interesting cases known to us as "alternating personality." If there are two of these, we have *double* personality; if there are three or more of these personalities, we have a case of *multiple* personality. There are sometimes six or seven of these, and in one case it is reported that there were *ten*,—all in the same individual, all alternating with one another, all having their own prejudices, likes, dislikes, interests, points-of-view and knowledge of persons and things! Many such cases have been

cured by welding together several of these "selves"
by hypnotic suggestion, when the original man was
restored.

THE DIFFERENCES BETWEEN SPIRITS AND SUBCONSCIOUS "SELVES"

Now if this be true (and it has been proved to be true
by many well-authenticated cases) how distinguish these
"selves" from true spirits? This is a very complex
question, which cannot be fully answered in this place,
because we must understand, first of all, more of the
nature of the subconscious mind and its powers. But
one simple test can be applied, which is this. All these
personalities, or parts of selves, derive their knowledge
of men and things through the same source, namely
the *five senses*. None of them can possibly know any
fact which was not supplied to them through sight,
hearing, touch, etc.,—so that, if any of them manifest
supernormal knowledge, this proves to us at once, either
that some external intelligence is present, or that this
personality, whatever it may be, has acquired this knowledge in some occult manner,—by telepathy, clairvoyance,
etc. Which of these two interpretations is the correct
one I shall endeavour to answer in another place.

PERSONALITIES CREATED BY HYPNOTISM

It may seem incomprehensible to many how the human
mind can become split up or "dissociated" in this manner. They think that this is rather a far-fetched theory
and prefer to believe the simple theory of "Spiritism"
as applied even to the most simple facts. This might
be admissible but for the following consideration: We
can trace a gradual series of intermediate steps all the

way from normal states of mind to these dissociations. In day-dreaming and absent-mindedness we see the first of these steps. When we hypnotize a subject and suggest to him that he is Napoleon Bonaparte or Julius Cæsar, and he enacts the part with due gravity, we can hardly suppose that Napoleon Bonaparte or Julius Cæsar *really* returned to manifest through him! And should any be inclined to accept this view, it may be said that the hypnotic subject will just as easily carry out the suggestion that he is a lion or a bear or a bird flying in the air,—and no one, we imagine, would contend that a lion or a bear or a bird really manifested at such times! So, therefore, we see that one part of the mind may enact a little comedy by itself, without the knowledge of another part; and, from this simple fact to the most striking phenomena of the subconscious we can trace a definite chain of connection.

THE PICTURE-FORMING FACULTY OF THE MIND

One of the most striking powers which the subconscious mind possesses is its ability to reconstruct mental pictures or photographs of distant or imaginary persons. Your mind contains a whole picture-gallery of all your friends, which you see, as it were, in your "mind's eye." This is limited not only to your friends and relatives, but to heroes of books you have read, and even to imaginary personages. These pictures are not set and inert; but live and move; and we place the characters in various situations and cause them to move, act and talk as human beings would do. Thus, suppose your friend "A" or David Harum, or some imaginary personage, were thought by you to be on a journey. You would imagine them to be in various situations, and would picture to yourself precisely how they would act in each situation, and would put into their mouths arguments and conversations which they would carry on with those about them.

This faculty which the mind possesses is a very peculiar one, and its functions are technically known as "Spiritoid functions." They have a great bearing upon Spiritism.

HOW DREAM PERSONALITIES TALK

These phenomena show us how easy it is for the subconscious to imagine that various personages are present, carrying on a conversation with us, etc., whereas, as a matter of fact, they are not present at all, but were invented by us. If, therefore, at a séance, some exalted personage appears and claims to communicate, we must always assure ourselves first of all that this personage is not one of these semi-conscious or subconscious creations, and must make him give proof of his own identity.

This faculty of the mind is again seen in dreams. In dreams we create situations in a similar manner, and imagine that other personages are present talking to us. We have long debates and arguments with such personages, and sometimes they beat us out! So you see how important it is to be sure that the Intelligences which communicate at séances are not creations, but are really individualities, as they claim to be.

HOW TO DISTINGUISH TRUE FROM FALSE

"How are we to prove this and make this distinction?" you may say. The following is the first method:

All our knowledge, whether it is conscious or subconscious, is supposed to be obtained through the five senses. The subconscious is built upon the facts obtained by means of hearing, sight, touch, etc. Now, if the communicating Intelligence tells us many facts (as proof of its identity) which the mind of the medium *never knew*, we have fairly good proof of identity—or at least that the knowledge given was obtained by some supernormal means.

But, decisive proof is not yet obtained! We know that there are other methods of obtaining supernormal information, for instance, telepathy, clairvoyance, etc. Adding these powers to the subconscious faculties of the medium, we have often a difficult task to prove that the Intelligence which communicates with us is really the personage it claims to be. Repeated questions must be asked, absolute proof of identity must be insisted upon, and in this way only can we be sure that we have passed beyond the limitations of the subconscious mind of the medium, and that we are really obtaining messages direct from the Spirit World.

WHY AND HOW SPIRITS PROVE THEIR IDENTITY

This proof of identity is really the great problem and the first point to solve. Let me make this plain. Suppose that a cousin of yours had disappeared twelve years ago. One day you receive a call over the telephone and a voice says to you, "I am your cousin, so-and-so. I demand my share of your uncle's will!" Naturally, you would reply, "How do I know that you are so-and-so?" In daily life, it would be an easy matter to prove this; he could appear before you in his physical body and you could identify him,—more or less easily in most cases. But suppose he were so placed that he could *never* see you personally. In such a case, how could he prove to you that he really was the person in question?

He would have to relate to you a number of personal and detailed incidents in his past life, which he would be the only likely person to know; or, relate facts which only he and you knew; or, tell you things which you did not know, but which you afterwards found out to be correct. If you received a number of these replies, you would be right in concluding that he really was the person at the "other end of the line"; and this is the way in which spirits prove to us their identity. Until they do so, we can never be sure that the teachings they give are correct. If they succeed in proving their identity we may then accept their word as to the conditions of the next life, and other matters, since they were always truthful people in this life, and we have no reason to suppose that they are other than truthful now.

HOW WE OBTAIN SPIRIT MESSAGES THROUGH THE SUBCONSCIOUS

The subconscious is the channel through which we obtain spirit messages, in nearly all cases. That is, they come *through*, or by means of, the subconscious mind, and it, therefore, assists the "spirits" to communicate. The spirit can manipulate or act upon the subconscious while it cannot readily affect the conscious mind. This we can see ourselves in the following way: Many of us have noticed that just as we are dropping off to sleep, a forgotten memory has flashed into the mind. It could not find its way to our conscious attention while the latter was busy with the day's activities, but as soon as the conscious mind became passive, then the subconscious had the power to send up this memory or message of warning. It is the same in the case of "spirits" who communicate. They are only enabled to do so when the conscious mind is in abeyance,—quieted, or abstracted, more or less completely, as in a trance. Then the spirit is enabled to act upon the subconscious mind of the medium and, *through* it, to reach us still in the body. The subconscious is, therefore, the true medium or *vehicle* for the manifestation of discarnate spirits, and this will become more apparent when we come to consider the phenomena of trance, which will be dealt with more fully in a later chapter.

CHAPTER V

THE SPIRIT WORLD

ORTHODOX theology has always taught us that, when we "die," we pass into either one of two places: Heaven or Hell. The Catholic church introduces a third intermediary state, Purgatory; and when in this state, souls may be helped either by those who have passed over or by the prayers of the living. It will thus be seen that, in this respect at least, the Catholic church approaches nearer than any other religion the doctrines of Spiritualism!

Information regarding the spirit world has come to us in various ways. Seers or clairvoyants have gone on "spiritual excursions" into the spiritual world, and have told us, on coming back, what they have remembered of their clairvoyant visions. Moses, St. John, Swedenborg, Andrew Jackson Davis and others were seers of this type.

On the other hand, we have the direct statements of "spirits" who have come back and related to us the precise conditions existing in the next world. From both these sources spiritualists have succeeded in constructing a fairly complete representative picture of the next life and its various activities. I propose here to give a rapid and more or less dogmatic résumé of these teachings—without fully endorsing them myself, but merely asking the reader to form his own opinion concerning them.

APPARENT CONTRADICTIONS

There are various contradictory teachings, regarding the future state, which have been given us from time

47

to time in the past, and it has been held by many that, because of these contradictions none of them can be trusted; consequently none of the descriptions can be true! Thus, "spirits" who return through many French mediums declare that reincarnation is a fact, while those who return through English and American mediums declare that it is not a fact; etc. How are we to account for these discrepancies? As this is a stumbling block to many spiritualists, the reason for these contradictions must be given at once.

The answer is, as a matter of fact, simple enough. "Spirits" tell us that, after death, they are by no means omniscient. On the contrary, they enter the next life, as before said, carrying with them all their prejudices, beliefs and pre-conceived opinions. Now, this being the case, we can see that a spirit who, when alive, believed in reincarnation would, after death, continue to believe in it, and he would naturally gather round him or drift into the company of those who also believed in it. In returning through a medium, therefore, he would state dogmatically that reincarnation was true! He would merely express *his own belief,* which might or might not be true. On many points of this nature we have no absolute means of arriving at the truth. "Spirits" tell us their *convictions,* their beliefs, and these are founded on observation, or the wisdom of those spirits who have progressed greatly since their departure from earth.

THE DOCTRINE OF "ZONES" AND "SPHERES"

Many "spirits" teach us that the spirit world is composed of a number of "zones" and "spheres," one upon the other. Some have stated that there are thirty-two such zones, others sixteen, but the greater number have declared that there are but seven,—beginning with the one nearest the earth, in which are earthbound spirits, and progressing gradually until they are inhabited by more and more spiritualized beings. These Zones are said to exist one beyond the other, like the various layers of an onion.

On the other hand, others tell us that there are no such things as zones or spheres, but that Heaven or Hell are merely mental states, and that the various degrees of spiritual perfection represent the different zones. They do not occupy space, that is; they exist purely in the mind of the individual. Yet, perhaps, these two may be but two aspects of a single truth! It is only natural to suppose that those of similar interests would gravitate together just as they do in this life, and shun the society of others less evolved than themselves (unless they chose voluntarily to help them as occasion arose).

This being the case, those more advanced spiritually would congregate in certain places, and those less advanced would gather together in other places; so that, although the zones would not exist as physical spheres, shut off from each other by *physical* barriers, as many believe, yet they exist practically,—the barrier being a mental or spiritual one.

CONDITIONS AND OCCUPATIONS IN THE SPIRIT WORLD

Spiritualism teaches that the next life is a busy one; that we continue our pursuits, activities and interests just as we do here, only under more favourable conditions. Evolution reigns supreme,—just as it does in this world. This is only natural and rational and what we should expect. It is a gradual continuation and process of ad-

vancement. The next world is said to be more or less
a duplicate of this one. Those who are interested in
learning may attend lectures or schools of instruction,
may read, write, compose, paint, play, etc., just as they
do here. The scenery is more or less similar to the
scenery on this earth, although more beautiful and perfect in every respect. We are told that children never
enter the lower spheres; nor are there any flowers in
these spheres, they are found only in the higher spheres
or more advanced stages. These spheres can influence
one another more or less directly to a great extent, and
particularly the higher spheres can exert a helpful influence on the lower ones. For this reason progress is
always possible for a Spirit who desires it. He can
secure assistance from those who are more advanced
than he is in the spiritual world. His progress would,
therefore, be rapid; and it all depends upon individual
effort how rapid this will be. The sooner a spirit realizes
his own possibilities, and the fact that his own future
happiness or unhappiness depends upon himself, the
more rapidly will he advance.

THE SPIRIT BODY

"Spirits" tell us that we inhabit, in the next life,
a body similar to the material body,—but representing
the glow of youth in its strength and purity. "The
spirit of man is ever young," and that being so, it
assumes that rejuvenated outward appearance, upon
entering the new life. This etheric body is incapable
of fatigue, and is fed by the magnetic and spiritual
forces which surround it in that sphere. Children,
entering the new life, gradually grow to maturity, though
more rapidly than they do on this earth, because greater

advantages are offered them, and progress is consequently swifter. At the age of greatest mental and spiritual maturity they cease growing, and thenceforward remain in that perfected condition.

ON ENTERING THE SPIRIT WORLD

Upon entering the next life, the human spirit is met by friends or relatives who have before passed over and who are drawn, by natural magnetic attraction and sympathetic interest to those who have just entered the "Spirit World." When the spirit enters the next life, it undergoes in a way a "new birth," and is for some time bewildered. This is only natural after the shock and wrench of death. When we have had an accident in this life, and have been knocked unconscious, the process of regaining consciousness is peculiar. When such a man opens his eyes, objects are presented to him vaguely, indistinctly. He would "see men as trees walking." Sounds would be heard but faintly. There would be a vague jumble of noises, and no definite and articulate sounds would be recognized at first,— until consciousness was more fully restored. Thoughts would be scattered, incoherent, and only the strongest stimuli would focus the attention on any definite object for longer than a few moments at a time.

When a man dies, the departure of the soul from the body must be as great a strain upon the surviving consciousness as any accident could be, especially in cases of sudden death, suicide and in those cases where the patient is said to "die hard." Of course, after a little time, the spirit survives the initial shock, and soon becomes adjusted to the new environment and condition; and this fact would account for the bewilderment and confusion which many spirits seem to experience upon their entering into the next life. It is only natural, and what we should expect.

SEX IN THE SPIRIT WORLD

Many have asked whether the distinction of sex is maintained in the next life; whether man continues to be man and woman woman. Here, again, many different opinions have been expressed by those who have passed over, but the majority seem to contend that the distinction between male and female is fundamental,—mentally and spiritually no less than physically, and for this reason they are destined to be more or less different for all time. This does not mean, as many think, that woman is there (as she is here, too often) in a condition of subservience or inferiority. On the other hand, she is man's equal in many particulars; in some ways inferior to him, and in some ways superior. It is a question of differing viewpoint and constitution. Each may attain perfection and ultimate complete happiness in their own particular way,—just as every individual here must obtain it in his own way.

As to the relations of the sexes in the next life, the teaching of the highest spirits is that there is love, harmony, sympathy, co-operation and a mental and spiritual blending together of their natures which corresponds to physical love on this plane.

"Earth-bound spirits" in the lowest plane are said to be unable to get away from the "atmosphere" and "magnetic attraction" of this earth, and do not care to, even if they could. They are the cause of much of the trouble which mediums experience, often causing obsession by delivering false or lying messages.

There seems to be a law which permits "spirits" from the higher zones to descend into the lower zones, but the reverse of this does not take place. Thus, there are good or spiritual influences always playing upon the lower spheres from the higher spheres, and progress is thus rendered easy to those who care to take advantage of their opportunities.

WHERE AND HOW SPIRITS LIVE

Many of the descriptions which have been given to us indicate that spirits inhabit houses or "mansions" very similar to our own, and that the scenery of the spirit-land is also similar to that of the earth plane, —only more beautiful. "Garments" of variegated colours are said to be worn, as well as ornaments for those who care for them. The occupations of spirits are many and varied. Time is not spent in the spiritual spheres, as many imagine, in idleness or in religious devotions.

HOW SPIRITS TALK

The "spirit-body" in the "spirit-heaven" is thus as material to them as our world, only it exists on a different plane of activity, and vibrates at a different rate of activity from ours,—hence it is invisible to us, as we are usually invisible to them, and it requires clairvoyance on the part of spirits to perceive the material world, just as it does on the part of mortals to perceive the spiritual world. Conversation between "spirits" is carried on by a species of thought-exchange or telepathy, though the conversation appears perfectly natural and as though delivered by means of mouth, as it is with us. We can form some idea as to how natural this would be from our *dreams*,—when the exchange of thought is purely mental, yet the words spoken to each other by the dream-figures seem as natural and as sonorous as our usual conversation.

INSANITY AND SPIRITS

There are, strictly speaking, no insane spirits, it is said, except in the earth sphere, and these, previous to their insanity, were degraded spiritually and morally. They frequently continue in some degree insane for a long period of time, their spiritual condition not being favourable to their restoration, and here they are often attracted to mortals with like tendencies whom they obsess and through whom they ventilate their own disordered fancies and even impel them to acts of violence. However, as much insanity is caused by disorders of the *links* between body and mind, and, as these are all severed at the moment of death, the mind is usually normal and sound as soon as it enters the spirit world, and in any case it recovers very rapidly upon its entrance into that realm.

HOW SPIRITS TRAVEL

"Spirits" are said to possess the ability to move from place to place with extreme rapidity,—the fact, "as quick as thought," as the saying is. It is as easy for one to imagine oneself in China or in England as it is to imagine oneself in Brooklyn, if one is living in New York. The one process takes no longer than the other, and, as you are (in the spirit world) where your thoughts and interests are, you may readily perceive that it takes you no longer to reach one place than it does another. However incredible this may seem, at first sight, it is quite intelligible when we remember the rapidity with which wireless messages travel,—flying through space at the speed of light (186,000 miles a second) this would carry these waves nearly 7½ times around the world *in one second*, and it has been experimentally proven that these electric waves do travel at that rate. Such being the case, we can at least conceive that thought can travel at as quick a rate, however inconceivable it may appear to our reason.

GOOD AND EVIL SPIRITS

We have heard much of obsessing and lying spirits, of evil spirits and those who work harm, but we must remember that there are spirits of quite another character in the "Heavens," who are said to protect and guard us, give us wise counsel and advice, and are, in fact, veritable "Guardian Angels." Their duty is to impress our minds, and by this means to instruct and guide us, to instil good thoughts and resolves, admonish us of our faults, reprove us when we go astray and assist in the development of special talents. They do not interfere directly in the physical world, but impress our minds, influencing them in this way or in that.

THE DOCTRINE OF "CORRESPONDENCES"

There is said to be a definite agreement or correspondence between the material and spiritual order of things. What we perceive as a tree in this world is only the outward manifestation of the real spiritual tree lying within it, and this is true of all physical manifestations and facts which we see in nature. Every physical body has a corresponding spiritual body behind it, and this fact gave rise to the famous doctrine of "Correspondences" elaborated by Swedenborg. This correspondence throws a little light on the bewildering fact that spirits often speak of spirit-gold, spirit-marble, spirit-houses, spirit-books, etc., as if they were tangible realities,—not, of course, that these are sublimations of corresponding objects of earth, existent throughout but different as to material, yet sufficiently alike to be called by the same name. In other words, these spirit-objects are expressed in a different vehicle of the nature which is to us, externalized as gold, marble, etc.

We must endeavour to realize the reality of the spiritual world, which we have been unaccustomed to think of as in any way substantial owing to the teachings of theology.

THE DIFFICULTY OF DESCRIBING THE SPIRIT-WORLD

It is impossible to express things psychic adequately in direct language for the simple reason that our words are images drawn from material things and their effects. Immaterial things and the life beyond must, therefore, generally be described by symbols rather than by words, and these symbols (whether seen in vision or representing themselves to the mind in the normal state) partake less of the seer's idiosyncrasies than any direct language would do. This symbolism is often carried to a high place in interpretation—so much so that the original is almost lost sight of. Of this, however, we shall speak at length in the chapter devoted to ''Symbolism.''

There is much evidence to show that spirits can create forms and objects by the mere exercise of their volition. They build up what appear to be solid objects by the use of their minds, and these objects are often mistaken by the spirits for realities. Thus, thought-forms may be created by a spirit intelligence, and this is a fact which many spiritualists have overlooked,—though it is an important one as I shall endeavour to show later.

DARKNESS AND LIGHT

Wrong and evil in some ways seem connected with darkness. "Unhappy spirits" always complain that they can find no "light," but, as they progress, the Darkness seems to lift and Light begins to dawn. This does not mean that they emerge from a material darkness into a material light, but go through a process of psychic evolution, which would, in their own minds, correspond to this. The quickest way for an unhappy spirit to progress towards the light is for it to help and comfort or assist another in like condition. Unfortunately, they are very often ignorant of this, but fortunately many spiritualists have done a great deal of good in the séance room, etc., by giving this knowledge to "spirits" of a low order. Many of the "spirits" who have passed over, being nearer earth than "Heaven," soon after their transition are more easily reached by the living than by other spirits,—so far as comfort and advice and assistance are concerned,—and for this reason prayers of the living are often of great help to those who have recently passed over and are extremely earth-bound by reason of their mental and moral characteristics. Ordinary advice and assistance may also be given to these spirits at a séance.

VISITS TO THE SPIRIT WORLD

The spirit world can occasionally be visited, it is said, by the spirit of the sleeper or the somnambulist, and the deeper the sleep, the more separated from the body

is the spirit—until, in deep trance, the spirit is sometimes entirely withdrawn. In deep sleep, also, the spirit
occasionally goes on clairvoyant excursions, and comes
back to its normal body,—remembering much that it has
seen in the spiritual realms. In the state of "ecstasy"
these voyages are often made, and the seer will retain
a certain amount of consciousness of this earth and be
able to dictate to those about him his impressions while
visiting the spiritual world, and while seeing more or
less clearly what is happening there.

"Spirits" are said to exercise free-will and have far
more liberty of choice in the next world than they do
here—where they are bound by habit and tradition no
less than by mental and physical obstructions and difficulties. The psychic gifts of spirits are far more highly
developed than they were when on this earth, and they
are frequently capable of exercising the faculty of fore-
knowledge or prevision as well as other supernormal
powers,—such as telepathy, clairvoyance and clairaudience.

INFINITE INTELLIGENCE

They are also able to perceive the general Plan of
Nature far more thoroughly and effectively than we,
because they have, so to speak, a greater mental *grasp* of
the Universe in its entirety; and many spirits who have
died while disbelieving in an Infinite Intelligence have,
as time progressed, shown that they have more or less
changed their viewpoint and now are more definitely
religious than they were before. As Dr. Crowell says,
"I have constantly been impressed with the numerous
proofs of the creative and sustaining power of Deity,
and step by step I have been led to undoubtingly believe that He, though not in human form, is everywhere

present,—the Creator, Preserver and Controller of all things,—literally God, in the most comprehensive sense of the term, with whom all wisdom and power and Infinite Love extends to all his creatures. This is the effect of these investigations upon my mind, and I am disposed to believe that similar and more extended researches by others in the future will lead all true, earnest spiritualists to the same belief, and thus modern Spiritualism will be stamped with the higher polity of true religion with a correct, though necessarily limited, conception of God's character, and of his relations to us and of ours to him.''

SHALL WE ''SEE GOD''?

However this may be, it is claimed that "spirits" for some time after transition at least do not definitely know anything more about the nature or extent of this Infinite Intelligence than we do. They do not pass *directly* into the presence of any Deity, as theology tells us. Questioned on this fact, they reply, ''I do not know!'' However, as they progress in spiritual perception and understanding, they gradually perceive that the Universe, instead of being a Chaos due to chance, is orderly and systematic, and governed by a Supreme or Infinite Intelligence which is the Guiding Principle involved, and that it would only be logical to believe that such an Intelligence necessarily existed.

THE SPIRIT WORLD THE SOURCE OF ENERGY

The spiritual world is the source of all energy. Even in this life our energy is derived from some spiritual source. The nature of life is as yet unknown, and there is every indication that it is due to some spiritual influx acting

upon and through the material world. One proof of
this is that, during the hours of sleep, when the body
is resting and passive, the nervous or spiritual energy
is revived, the body is recharged, as it were, in the same
way as a storage-battery might be recharged with electric energy. This process does not depend upon any
material condition,—for sleep can often revive us instantly, as many can attest. In moments of extreme
exhaustion, the head may drop to the breast for a fraction
of a second, and a moment later consciousness be re-
gained, yet, in that moment of time, some complete
spiritual revivification has taken place. The energy of
the body seems to have been recharged or replenished,
and new energy infused from some spiritual source in a
manner which would be quite inexplicable, were we to
depend upon the ordinary teachings of science to explain
such facts.

The phenomena and teachings of Spiritualism alike
constitute a great solace and comfort to many souls
in distress and sorrow. The proof that death does not
end all and that the individual human spirit continues
to exist as an entity and in precisely the same form
as it is now, is a great comfort to the majority of
persons. In this way, the teachings of Spiritualism
are a solace to those who accept them. To those who
not only believe, but are enabled to obtain some of the
varied phenomena, this assurance and consolation is
doubly true.

THE DIFFERENT KINDS OF "SPIRITUAL GIFTS"

There are many "spiritual gifts," as St. Paul says, in
his message to the Corinthians. He wrote, "Now there
are diversities of gifts, but the same Spirit, And there

are differences of administrations, but the same Lord, and there are diversities of operations, but it is the same God which worketh all in all, But the manifestation of the Spirit is given to every man to profit withal. For to one is given by the Spirit the word of wisdom, to another the word of knowledge by the same Spirit, to another faith by the same Spirit, to another the gift of healing by the same Spirit, to another the working of miracles, to another prophecy, *to another discerning of spirits,* to another diverse kinds of tongues, to another the interpretation of tongues; but all these worketh that one and the self-same Spirit dividing to every man severally as he will.   (1. Corinthians, Chap. XII.)

How any one can disbelieve in spirit communication on the ground that it is contrary to Bible teachings, after the above passage, it is hard to comprehend,—since here are a large number of spirit manifestations clearly outlined and stated by the Apostle to be manifestations of the Divine Spirit!

CHAPTER VI

THE HEALTH OF MEDIUMS AND PSYCHICS

THE health—bodily, mental and spiritual—of mediums is a very important factor in all mediumistic and psychic development,—far more so than is usually realized. In the first place, we have a certain amount of bodily energy in order to accomplish anything we desire in life, and this energy comes largely from physical health. Mediums have found to their cost that the production of phenomena (especially of the physical order) is at times a very exhausting process, and unless they keep themselves in good bodily health, they discover that they become run down and nervously exhausted, in which case they render themselves subject to insomnia, depressing mental emotions, and, if this gets worse, to obsession and even greater dangers and difficulties. It is very important, therefore, for all mediums to keep up their physical health.

A HEALTHY BRAIN

The mind of man depends largely upon the condition of his brain and, if this is not rested, freshened and supplied in abundance with rich, healthy blood, his mental life suffers in consequence, for we know that any poisonous substance, mixed with the blood, immediately affects the mind by circulating through the delicate substance of the brain. The tiny nerve-cells, all over the body, which are the storehouses of energy, may be compared to a number of tiny cups, which we fill with energy every night during sleep and more or less empty every day. Our duty is to keep these little cups brim full, and if we allow them to become too emptied, so that nothing is left, we run into danger of nervous exhaustion, neurasthenia, etc. The first thing which the medium must pay attention to is, therefore, the state of his physical health: and the following rules will be found helpful by all those who wish to attain this condition.

DEEP BREATHING EXERCISES

In the first place, a certain number of deep-breathing exercises should be taken every day. These serve to keep the lungs active and to massage the internal organs. But deep breathing exercises have a more potent and far-reaching effect than this. There is a peculiar life-giving property in fresh air, and if we do not breathe this fully, we never *live* as completely and receive as large a supply of the vital and magnetic currents of the universe as we otherwise would. If any one doubts this he has but to stand erect and take half a dozen deep breathing exercises, as directed below, and he will feel energized from top to toe. The way to take these breathing exercises so as to get the best results is as follows:

HOW TO BREATHE

1. Stand before an open window or out of doors,— free from all restrictive clothing. Before beginning, exhale forcibly, bending the body forward, and relaxing the muscles. Place both open hands over the abdomen. Now breathe as deeply as possible *against* these hands, expanding the abdomen as much as possible, without allowing the chest or ribs to expand in the least. In

other words, breathe with abdomen only. After you
have done this five or six times, place both your hands
against your ribs on either side. Now breathe in deeply,
pressing out the ribs, but without allowing either the
abdomen or the upper chest to expand. After you have
done this five or six times, place your hands on the
upper chest, just below the neck, and breathe with this
portion of the lungs, without allowing either the ribs
or the abdomen to expand. At first you will find it
very difficult to control your breathing, limiting it to
these parts of the lungs: but this will come with practice,
and it will be shown later in Chapter XLI how important
these breathing exercises are, when the psychic side of
the breathing exercises is understood.

THE PSYCHIC (COMPLETE) BREATH

After you have mastered these three separate steps,
you will be enabled to take what is known as a "complete
breath,"—that is, one which expands first the abdomen,
then the ribs, then the upper chest. You should, by
this time, have such control over your breathing that
you are enabled to do this in three distinct stages, or
merge them together into one, as you wish. In all these
breathing exercises, the back of the nasal passage should
be relaxed, and you should breathe through the nose
(never the mouth) as though you were smelling a flower.
If you do this and relax inwardly, you will find that the
air strikes the back of the throat before it is felt at all,
and you will never notice the air *in* the nose itself.
Practice this every day until you become proficient.
The best way to insure this is to close the lips, while keeping the teeth separated; then throw down the under portion of the jaw.

DEVELOPING EXERCISES

2. A certain amount of exercise should be taken each day. The particular character of exercise which will be found beneficial for the maintenance of health, also for the development of psychic and mediumistic gifts, are those which develop the vitality of the inner organs, about the waist line. Bending exercises of all kinds are especially useful. Large muscles are not required for good health, but energy and endurance are. The following four exercises will be found very helpful, in this connection.

(a) Stand erect, raising both arms over the head as far as possible. Raise yourself on your toes and, at the same time, stretch upwards with the finger-tips as far as you can, as though trying to lengthen yourself.

(b) Stand as before, arms raised over the head: now bend forward and try to touch the floor with your finger-tips without bending the knees. Again raise yourself to a standing position very rapidly. This is a well-known but very useful exercise.

(c) Stand as in exercise (b) and bend the body sideways from the waist as far as possible—first to the left, then to the right. Make this motion as rapid as you can.

(d) Stand as before and bend slowly, trying to touch the floor with the fingers. As you do this, take in a deep breath. The purpose of this exercise is to compress the liver from above and below at the same time, and this massage will prove very helpful.

HEALTH HINTS

Other points to be observed in maintaining good health are the following:

1. Eat as little red meat as you can, since this is acknowledged by all to retard psychic development.

2. Eat a certain amount of fruit every day, not in addition to other food, but in place of it. Acid fruits are particularly beneficial in nearly all cases.

3. Drink at least a quart of water each day.

4. Accustom the body to cool baths. It is best to begin these in the summer-time and continue them in the winter.

5. Wear as little clothing as you can, consistent with warmth. The skin breathes as well as the lungs, and free circulation of air on the surface is essential.

THE POWERS OF THE MIND

We now come to the *mental* factor. Few realize how important this is, in the development of psychic gifts. If the mind be depressed, worried, scattered and unable to concentrate upon any definite thing, good results can hardly be hoped for in the way of psychic development! Many psychics can obtain good results for individual sitters, but as soon as they make a public appearance they fail more or less completely. We can hardly doubt that the reason for this is their apprehension for the results, fear that they will not succeed, etc. This prevents all free communication: it shuts the doors of the soul, as it were, against any outside influences. In order to be receptive and sensitive, we must have a free mind, and give ourselves up wholly to those forces and vibrations which play upon us. If you watch yourself you will find that your body tends to contract all over as soon as you think certain thoughts or experience certain emotions, such as jealousy, hatred, envy, etc. On the other hand, as soon as you send out thoughts of friendship, love, sympathy, etc., you find that your whole being expands and relaxes. If this is true of the muscles of the body, how much more true is it of the "muscles of the soul," if I may so express it! "The imagination" it has been said "is the lungs of the spiritual life," and in order to have free play, they must be unrestricted, just as our physical lungs are. The essence of psychic development is this complete surrender and quiescence, and until this is insured, full development can hardly be expected. There is such a thing as "spiritual contraction." We have all heard of the man with the "ingrowing conscience." This means, simply, that this man is dwarfed, contracted and unsympathetic in his attitude to all that he meets. "Gentleness and cheerfulness," said Robert Louis Stevenson, "are the perfect duties"; and we cannot do better than advise the medium to follow this motto in his daily life.

PSYCHIC CONTAGION

These influences which are harmful in ourselves are harmful when experienced in others, and they are contagious to a remarkable degree. All experienced spiritualists know that a medium is liable to "take on" the conditions of a spirit or of another person, when in a sensitive state, and this is true of his mental and spiritual life as well as his physical health. We can acquire the other's irritable disposition, his sourness and lack of balance, for the time being, just as easily as we can acquire other symptoms; and unless this is recognized and the medium takes care to throw off these influences, they are liable to remain with him more or less and influence him—just as we sometimes experience the after influence of a bad dream in the day-time.

HOW TO CHOOSE A GOOD DEVELOPING MEDIUM

The practical conclusion to be drawn from all this is that it is very dangerous to the mental and moral health of a psychic to develop under the guidance of a medium who is mentally, morally, physically or spiritually ill—for these conditions will possibly sooner or later be "taken on," and they are liable to influence the medium to his own detriment. Be most careful, therefore, in selecting the psychic under whom you develop, for your own future progress and happiness will depend largely upon that.

CHAPTER VII

SELF AND SOUL CULTURE

"Know thyself" was the mandate of the Delphic Oracle! Before man can undertake to govern and control external forces, he must learn to control those within himself, for only by doing this can success be attained.

Man utilizes his mind as he would a tool every day of his life. The better we understand our tools the better workmen we are. Hence he who would succeed must understand the workings of his own nature.

THE COSMIC CURRENTS

First of all we are told that there are Cosmic currents playing to and fro in the world, contradictory currents or streams of thought into which we are liable to enter unconsciously, even against our will. Some of these currents are beneficial, others are harmful. Some natures are strong to stem the tide and achieve success against the greatest obstacles; others can extricate themselves but partially, others do not do so at all. For this reason we have the successes and the failures in life. It depends partly upon outside influences, partly upon ourselves. The first we cannot control, except indirectly through ourselves.

HOW TO MAKE A SUCCESS OF YOUR LIFE

Here is the explanation of a great fallacy which many people make. They imagine that they can by their own will mould circumstances to suit themselves. This is only partly true. Let me explain. We must not turn our power of mind upon others, we must turn it upon ourselves in such a way that it will make us stronger, more positive, more capable and more efficient; and as we develop in this manner success will come of itself. The way to control circumstances is to control the forces within yourself, to make a greater man of yourself, and as you become greater and more competent, you will naturally gravitate into better circumstances. We should remember that "like attracts like." For, as Dr. Larson says "those people who fail and who continue to fail all along the line, fail because the power of their minds is either in a habitual negative state, or is always misdirected. If the power of mind is not working positively and constructively for a certain goal, you are not going to succeed. If your mind is not positive it is negative, and negative minds float with the stream. We must remember that we are in the midst of all kinds of circumstances, some of which are *for* us and some of which are *against* us, and we will either have to make our own way, or drift, and if we drift we go wherever the stream goes. But most of the streams of human life are found to flow into the world of the ordinary and the inferior. Therefore if you drift you will drift with the inferior, and your goal will be failure."

THE THREE LAWS OF SUCCESS

In order to achieve mental and spiritual success (and the same rule applies also to worldly success) three rules must be observed which are of prime importance: The first is that you must have in your own mind a *clear conception* of what you want. If you have not any definite goal in view, you cannot expect to achieve any great success, because you will be constantly wasting your energy in byways, without directing it all towards one certain point. The second is: You must make your *thinking* positive and not negative. This does not mean that you must grind your teeth, frown and try to dominate every one you meet. It means that you possess a calm self-assurance and the inner conviction and certainty that you will succeed. Physically this state of things may be felt in a full, firm sensation throughout the nervous system. The third rule is: All your *thinking* must be *constructive*, that is, built about the goal or object you have in mind. If you spend only a fraction of your energy of thought in any one direction, you cannot expect to progress very far in that line. The runner who tries at the same time to work out a mathematical problem in his head will not be first in the race! Constructive thinking means that you must consistently and continually think of and about what you wish to accomplish.

The sooner you learn to do this the sooner will success be yours. Obstacles in life present great difficulties. Up to a certain point they may be looked upon as helps to character and progress and the more these are overcome the stronger will your character ultimately be. At the same time this may be overdone, and there is such a thing as "Kicking against the pricks."

FLOWING WITH THE TIDE

If you are striving your best (and every man knows in his heart when he is doing his best) to accomplish a certain thing, and more and more difficulties seem to multiply the further you progress, you may, under certain conditions, assume that it is not meant for you at this particular time to do this particular thing, and you may shortly look back and see how you were prevented from undertaking something that might have proven disastrous. In this way it is possible to float with these "Currents" instead of stemming them to advantage. Mrs. Towne tells us that she, at one period of her life, could do nothing on account of her desire to rest and sleep. She determined that she would give this full play; she went to bed and stayed there for fourteen days and nights! At the end of that time she felt that, at last, she had had enough rest, and thenceforward work became a joy instead of a burden. It proved to be the turning-point in her life.

ACTING FOR OUR ULTIMATE GOOD

There are, therefore, Cosmic Currents swaying to and fro, flowing back and forth throughout the Psychic Universe, and the more we can "sense" or become receptive to these currents, the more will our life be guided and directed for us by an Intelligent Control, greater than our own. We all think that we know exactly what we want to do, and what is best for us,—yet this is not always the case! To a mind vaster and more inclusive than ours the very opposite of this may seem better for our ultimate good. For example, a dog has to have a tooth extracted. The painful operation of removing the tooth is all that the dog can see. To him it is all painful, nothing beneficial. To us, on the contrary, who see not only what a dog sees, but more, it is clear that the dog will eventually be better for the removal of his tooth—though it is a painful experience. Applied to ourselves, it is most probably true that our painful experiences in life can be interpreted in a similar manner,

and that many of them, could we see them in that light,
are for our ultimate good.

HOW THESE LAWS APPLY TO PSYCHIC DEVELOPMENT

Now let us apply what we have learned to psychic
development, and the cultivation of mediumistic gifts.
We have found that there are magnetic and spiritual
forces playing upon us from different directions here and
there all over the world. Some of these are for our
own good, others are not. We must learn to become
sensitive to those currents which are beneficial to us,
and shut out those which are not. How are we to do
this?

In the first place it is necessary for the student who
really desires to obtain this guidance to make certain
renunciations or sacrifices. He cannot be "in the
world," and at the same time receive this spiritual
perfection. One cannot both eat one's cake and have
it! So you must make up your mind just what you wish
to do. Many mediums unfortunately do not develop
along this line. The cultivation of the spiritual self
is not altogether the same thing as the cultivation of
the psychic self,—obtaining psychic phenomena. The
one great reproach which has been made against many
mediums and spiritualists (it must be admitted, with
some justice), is that "spiritualists are everything but
spiritual!" Doubtless this is not true of spiritualists
any more than the followers of any other religious faith.
Human nature is weak, and we all fall from grace. But
we are now only talking of those who sincerely desire
personal spiritual enlightenment, and who are willing
to make some sacrifices in order to obtain it. To those
who are anxious to follow this path we would say that

it is unwise to give too full directions thus early in your
development. This is a question which will be discussed
more fully in Chapter XLI. For the present a few
practical points may be helpful both in your daily life
and your psychic unfoldment.

RULES TO FOLLOW

1. As is so often insisted upon, the health must be
maintained. If this is not done, you render yourself
liable to nervous exhaustion and, through this, to obsession.

2. Your clear common-sense and interest in the things
of the world must, to a certain extent, be kept. Otherwise the judgment will become unbalanced.

3. Cultivate sympathy, harmony, interest in your fel-
low-beings.

4. Cultivate your own sensitiveness along ordinary
psychic lines, by various special exercises. When you
obtain a certain number of psychic phenomena in this
way, you will be far more receptive than you were before.

5. Cultivate at all times what may be called *a listening
attitude of the soul*. This is particularly important and
practically valuable. When you are in doubt upon any
question, retire to a quiet room and ask your own higher
self what is the best thing for you to do. At first these
replies will be very vague and indistinct, but as you
progress in your development you will find that they
will become clearer and clearer, and you will soon get
definite and clearly formed replies in answer to this
mental questioning. As soon as you have progressed
thus far, you may be sure that you have begun to sense
the "Cosmic Currents" which flow about you; and when

once you have done this, it is, thenceforward, only a matter of personal development. This will be dealt with more fully in several of the chapters which follow.

CHAPTER VIII

THE CULTIVATION OF SPIRITUAL GIFTS

THERE are two ways of regarding any particular fact; the first is to observe it from without, the second is to experience it from within. If we look at an orange, we observe it from without, and we could never experience it from within, unless we *were* the orange. The only things that can experience sensations from within in this way are *minds*. Each mind can inwardly experience and see objectively its own sensations, and, so far as we know, it is the only thing in the world which can do so. All psychic experience is, therefore, inward or sensitive, and can never be felt by another person, but must be experienced by him in order that he should appreciate and understand the mental state you yourself are experiencing.

It is the same with psychic phenomena. If any one experiences any phenomena of this character, he can never impart this knowledge to others, except in a very roundabout way, and for these others to understand the phenomena they themselves must experience them. It is for this reason that it is so difficult for psychics to express and explain to outsiders the character of the sensations and phenomena they are experiencing. Everything being so largely symbolic, and our language being so poor in this direction, it is often very difficult for them to explain precisely what they mean.

DEVELOPING MEDIUMSHIP

We do not know as yet exactly what mediumship is. There is much evidence to show that it is very often hereditary and runs through three or four generations, just like any other gift. With some, mediumship appears in childhood and seems to be a very part of their constitution. The majority, however, develop it later on in life, as the result of coming into contact with mediums, or developing it within themselves by experiments. Some retain their mediumship throughout life, others experience it only for a few months, a few weeks, a few days, in some cases only a few seconds.

In some cases mediumship is terminated suddenly, in other cases it is gradually lost through a period of years. One who has at any time experienced mediumship can usually recall it by reason of its persistence no matter how long afterwards.

CONTROLLING PHENOMENA

In Mediumship, or when obtaining psychic phenomena of any character, we are as yet experimenting, as it were, with forces and laws as yet largely unknown— just as the early scientists experimented with electricity (indeed we do not *yet* know what electricity is). However at the present day we can control it perfectly, and it is to be hoped that the time will come when mediumship and all psychic phenomena can be controlled in a similar manner, even though we may never know the innermost essence of psychic power. If we could do that it would be, at any rate, on a "workable basis," so to speak.

MEDIUMISTIC EXERCISES

All mediumistic exercises develop this power to some extent, but in different directions. The following are a few of the methods which may be pursued in cultivating and developing the psychic self and the inner spiritual centre of our being, as distinct from purely psychical phenomena:

As before said, it is essential that we should understand and control ourselves before we endeavour to control outside forces.

Much may be learned through what is known as "introspection," that is the turning inward of the attention upon the inner self, instead of outward, upon the external world. If you close your eyes and do this and try to find out the nature of your true inner being, you will probably experience a peculiar sensation. You will find that, like Happiness, it continually eludes you, and that, when you think you have grasped your own Self, it is only a state of mind, which has since passed and is now only a memory!

Practice this introspection for a few minutes each day and before long you will be surprised at your development in this direction, for you will be enabled to come into far closer touch with yourself than formerly. The inner self will become illuminated, as it were.

MASTERING THE SELF

This practice will lead to the habit of escaping from our sense-perceptions, to which most of us are slaves. As you get away from these, and are enabled to withdraw more and more fully into your inner self, you will experience a sensation of reality and the ability to

perceive the truth of things in a manner hitherto undreamed of.

Truth exists; we do not perceive it for the simple
reason that the veil of sense is between it and us. Lift
this veil and you will perceive truth clearly, as in the
light of day.

This practice of acquiring greater mastery over self
will also put you more closely in touch with the great
magnetic power-currents of the universe, so that you
will never feel exhausted or in need of nervous energy,
—there being an unlimited supply of energy in this
universe. All we have to do is to learn to tap it,—
which we can do by these methods of psychic development,—and we can draw upon it in any quantity we
choose.

We will also be put in touch with higher conditions.

HOW SPIRIT UNITES US

With increased spiritual developmen. and spiritual
life, we will perceive that there is a universal brother-
hood of mankind and that nothing is really separated
from anything else; that we are not separated from
our neighbour, but that we are united in the great universal Infinite Intelligence which combines all. We may
compare ourselves to trees, in this respect. Each tree
is apparently, a separate being, whose leaves whisper to
one another, and whose branches sometimes touch in
the swaying of the evening breezes: but their roots are
sunk deep into the ground and are often intertwined
one with another, while the common earth unites them
all. In a similar way, we are united in the spiritual
universe, of which we form a part. Fundamentally,
psychically, we are united one with another.

MEDITATION

Meditation may be considered one of the methods by means of which we awaken the inner self and frequently awaken our spiritual or astral senses, so as to cause them to function on another plane.

At the same time, if this developing process is done properly, we build up walls of power about ourselves, which others will find it impossible to break through, by mental or hypnotic influence, even should they desire to do so. We cover ourselves with a sphere of energy through which nothing can pass, against our will.

THE POWER OF THOUGHT

All thoughts sent out by us into the universe have some definite purpose and have a certain effect, both upon ourselves and upon others.

"Thoughts are things!" We can create a thought as surely as we can create a house or a chair, and, once created there is no telling where this thought may stop or how lasting its action may be. If these thoughts are good, helpful and useful, they often return to us like boomerangs, with the added happiness and power which they have accumulated from others of a similar character in their flight through space. On the contrary, evil thoughts come back to us in the same way, and it will be found that they always return to their sender, with added power for evil or for good. See to it, therefore, that you only send out thoughts of the highest and best.

Some people—when they first realize this fact—are almost afraid at first to think at all, for fear of the *effects* their thoughts may have! But this is a great mistake. Expression is the first law of life. We must learn to express, *EXPRESS!* The chief outward difference between a living being and a corpse is that one can express itself and the other can not. Do not be afraid to express yourself fully and forcibly in any direction. Even the bodily expression of our feelings and emotions is quite justified. There is nothing to be ashamed of in conviction or in passion. It is the abuse of these which is detrimental.

THE USE OF THE WILL

In a similar way the power of the *will* may be used for good or for evil, as the case may be, and it has a great power in both directions—as the history of Occultism has shown us.

In the one case we have, as the result of the exercise of this power, various psychic phenomena, marvellous cures and all the varied accomplishments of this world. On the other hand, we have the phenomena of witchcraft, black magic, harmful absent treatment and crime.

It all depends into which channel we direct the energy of our will. The soul must learn to find and experience itself fully before it can consider itself thoroughly alive and a fully developed entity. After this realization has been accomplished, then, and then only, should we direct our attention to cultivating and directing the latent energies which we possess.

SELF-DEVELOPMENT ESSENTIAL

It is because of this fact that "Self and Soul Culture" is necessary, before psychic phenomena are cultivated to any great extent. We must learn to know ourselves, to preserve a just and careful balance of judgment, sympathy, understanding and intuition. If we do not possess these qualities, we shall never become mediums on the highest plane. On the contrary, we may draw to ourselves, while developing mediumship, harmful or lying intelligences, which we have attracted into our magnetic aura.

So I cannot too strongly advise and warn you, to practice these self-developing exercises before cultivating external psychic or mediumistic powers. Mediumship opens the doors to influence and powers over which we have little control, and we must be sure that, before the doors of the soul are swung back, we must be pre-pared to receive whoever enters, by reason of our own self-control and inner powers,—otherwise we may be unable to close the doors, when we wish to,—or the door of reason may become altogether unhinged!

In giving these warnings I do not wish to frighten the reader, since there is no necessity to become alarmed, if caution be exercised in this development. Only I wish to emphasize the necessity for this caution!

CHAPTER IX

PSYCHOMETRY

WHAT is Psychometry? Dr. J. Rhodes Buchanan says: "The word 'Psychometry,' coined in 1842 to express the character of a new science and art, is the most pregnant and important word that has been added to the English language, coined from the Greek (*psyche* —soul and *metron*—measure). It literally signifies 'Soul-Measuring.' In our modern use of the word, however, it means something a little different from this. A psychic who picks up an object and, in connection with it, gets certain psychic impressions, is said to 'psychometrize' the object, and this process is known as psychometry."

EXPERIMENTS

The famous Professor Denton, a mineralogist, whose wife possessed remarkable powers in this direction, conducted a number of experiments some of which are described as follows: He gave his wife a specimen from the carboniferous formation. Closing her eyes, she described swamps and trees with their tufted heads and scaly trunks, with the great frog-like animals that existed in that age. He got a specimen of the lava that flowed from the volcano in Hawaii in 1848. His sister, by its means, described a "boiling ocean," a cataract of golden lava that almost equalled Niagara in size. A small fragment of a Meteorite that fell at Painsville, Ohio, was given to his wife's mother, a sen-

sitive, who did not then believe in psychometry. This is
what she said: "I seem to be travelling away, away
through nothing, right forward. I see what looks like
stars and mist. I seem to be taken right *up*, the other
specimens take me *down*." His wife independently gave
a similar description, but saw it revolving and its tail
of sparks.

NOT DUE TO TELEPATHY

Prof. Denton took steps to prove that this was not
mind-reading, by wrapping the specimens in paper,
shaking them up in a hat, and allowing the sensitive to
pick out one and describe it, without any one knowing
which one it was. Among them was a fragment of brick
from ancient Rome, Antimony from Borneo, Silver from
Mexico, Basalt from Fingal's Cave. Each place was
described correctly by the sensitive in the most minute
detail.

These are but examples which could be multiplied,
did space permit. Nearly every one possesses a certain
amount of power in this direction, and it only needs
cultivation to bring it to light.

Before proceeding to the practical side of this question, a few words of explanation of the theory involved will doubtless be of interest to the student.

THE EXPLANATION

It has been said that every object possesses its own
peculiar psychic influence, fluid or aura, which may be
recognized by one sensitive enough to perceive it. Human beings may transfer a certain amount of this
"fluid" to objects, leaving them impressed with their
influence. We see this in the case of "magnetic cures,"
and in some cases of "haunted houses." In fact,—

as we shall see in Chapter XXVIII, devoted to that subject,—this is one of the theories which has been advanced
to explain haunted houses.

Objects which have been worn close to the skin, or
which have been brought into contact for a long time
with the magnetism of any particular person, seem to
retain a large share of this aura, and such objects may
readily be psychometrized—their aura may be read and
interpreted according to the ability of the psychic. We
often see demonstrations of this character given in
public. Again trance-mediums are very sensitive to
influences of this character, and if we place an object
which had belonged to some person who has recently
passed over into the hands of a good trance-medium,
he will frequently be enabled to get into contact with
that person, through the magnetism of the article in
question, and in that way information may be obtained
which otherwise could not have been secured.

HOW TO PRESERVE THE INFLUENCE

Articles of this character often lose their properties—
their "virtue" we might almost express it—by being
left around or exposed to the handling of others: and
for this reason it is best to keep such articles carefully
wrapped-up in thin rubber cloth which may be procured
from any drug store. In this way their properties are
preserved.

Just *what* this influence is, with which the articles
become impregnated, we are unable to say. Probably
it is a form of the vital force which animates the universe. Yet, even supposing that this could flow into
the object, and that the psychic could "sense" it, we
have yet to explain why it should be that this particular

vital energy should be enabled to arouse within the
psychic the flood of information he receives.

"AKASIC RECORDS"

Professor Draper has said: "A shadow never falls
upon a wall without leaving thereon a permanent trace
—a trace made visible by resorting to proper processes.
On the walls of private apartments, where we think the
eye of intrusion is altogether shut out, and our retire-
ment can never be profaned, there exist the records of
our acts, silhouettes of whatever we have done. It is
a crushing thought to whoever has committed secret
crime,—that the picture of his deed and the very echo
of his words may be seen and heard countless years
after he has gone the way of all flesh!"

There are certain analogies for this in the physical
world. If sunlight falls upon a sheet of paper and we
place upon it a key, the outline of this key will be marked
upon the paper and may be recovered years later by
suitable means.

If "thoughts are things," they doubtless impress our
surroundings in much the same way: and the objects
which we psychometrize are influenced by means of our
thoughts, and the human aura or fluid, so that they
retain them within it, and may be "read-back" by the
sensitive.

THE INTERPRETATION OF IMPRESSIONS RECEIVED

In all psychometry we must remember that the interpretation of the impressions received is largely symbolic,—just as the printed word of a book is symbolic
of the thought of the author, lying behind it.—So,
impressions stored within objects and "sensed" by the

psychic, must also be symbolic, and must be suitably interpreted by the psychometrist. Thus, when he places a geological specimen on his forehead and describes an "antediluvian monster," roaring and walking about, no one but a very shallow individual would imagine for a moment that the psychometrist was actually seeing the original! He simply got an impression of that era of the world's history, and symbolized it subconsciously in the form of this roaring monster.

In obtaining impressions from an object, we must endeavour to become as receptive and sensitive as possible. A few preliminary exercises will enable you to do this to much better advantage than you otherwise would be enabled to.

EXERCISES FOR DEVELOPING SENSITIVENESS

1. Cultivate the sensitiveness of your finger-tips. You may do this effectively by placing in a bowl water of the same temperature as the body. Now, close your eyes and place your finger-tips just above the surface of the water. Without looking, very gradually lower the finger-tips until they come into contact with the water. See whether you can tell when this is the case. You will be surprised to discover that, at first, you are quite unable to tell when you have touched the water!

2. Another good exercise is to take a pair of compasses and, opening them a quarter of an inch or so, touch the finger-tips with the two sharp points, the eyes being closed. See if you can tell how far apart these points are,—before looking at the compasses. In this way your fingers will acquire a sensitiveness of their own.

3. Learn to act upon first impressions. Do not hesi-

tate or be afraid to express exactly how you feel and the impression that comes to you,—no matter how "ridiculous" it may be. There is a useful saying which may help you in this respect. It is: "The first thought is the spirit's, the second is your own." So learn to act on first impressions, and put into execution immediately anything which comes to you.

4. Analyse your own sensations and emotions as best you can, after the first impression has been received, and see what you feel or experience within yourself. Then express this in words to the best of your ability. These emotions often express, in that form, facts which could not well be expressed in any other way, though they apparently have no connection with the object.

For example: If you are feeling a watch, and you get in connection with that watch the feeling of depression and pain in the throat, state this fully, since the person who owned the watch may have strangled himself in a fit of melancholy. In this way the emotions you perceive are fully in accord with the sensations which you receive from the object.

ITS PRACTICAL VALUE IN DAILY LIFE

The practice of psychometry will often enable you to tell the characteristics of another living person, and by this means you will be enabled to tell whether or not you will like such a person,—because you may be attracted or repelled by the psychic impressions you receive in connection with the object such a person has been wearing. In practical life information of this character is, at times, very useful.

In addition to all this, the cultivation of psychometry

is often useful in paving the way for the cultivation of other psychic phenomena, and will prove a useful Introduction to them.

CHAPTER X

THE HUMAN AURA

SURROUNDING every living body (and some non-living materials) there is a halo or "aura," which may be seen, under certain exceptional conditions. Clairvoyants have always contended that they could see this aura, surrounding human beings, but they were laughed at for their pains by the majority of scientists, who continued to disbelieve in its existence. About the middle of the 19th century, Baron von Reichenbach published a book on the aura, paying particular attention to the emanations which his sensitives had seen coming from crystals and the poles of horseshoe magnets. It is now known that both magnets and crystals give off a very noticeable aura, and this may be seen by any one, possessing even moderate psychic development, if they observe these objects when placed in a darkened room.

HOW STUDIED BY THE AID OF CHEMICAL SCREENS

Needless to say all this was disbelieved at the time, and it was not until 1911 that the existence of the aura was proved scientifically by means of mechanical and chemical means. Dr. Kilner, the electrician of St. Thomas' Hospital, London, then showed that it is possible for any one to see the aura, issuing from a living human being, by means of especially prepared glass slides, containing a chemical, named "Dicyanin."[1] The subject of the experiment is placed against a white or black cloth background, in a nearly darkened room, and must be at least partially nude, as the aura cannot be seen through the clothing. The investigator then

[1] Dicyanin is a coal-tar product or dye.

90

looks through one of the chemical screens at the daylight: then, closing his eyes, pulls down the blind, so as
to make the room nearly dark. In this light the figure
of the model can be seen only faintly, and if the subject
is looked at through the glass screen, the aura may be
seen by nearly any one, possessing good eyesight. In
this case the investigator does not have to be a clairvoyant, since the eyes are rendered susceptible to certain artificial light waves by means of the chemical screens.
Usually our eyes cannot perceive these waves.

In this way the sceptical world has been convinced
of the reality of the human aura, and it is now considered
a proved scientific fact.

THE THREE AURAS THUS DISCLOSED

The human aura or atmosphere consists of a number
of layers or strata one beyond the other, extending out
into space. By means of Dr. Kilner's chemical screens
three of these divisions may be clearly perceived.

First, what is called the *"etheric double."* This is
seen like a dark line, slightly greyish in colour, which
extends over the whole surface of the body, conforming
exactly to its shape. Doubtless this is one manifestation of the double or etheric body.

Beyond this extends the *"inner aura,"* which is usually two or three inches broad. It conforms to the contour of the body throughout and is more or less coloured
by the health of the individual and by the mental or
emotional states, which may be present at that time.

Beyond this again is the *"outer aura,"* beginning
where the inner aura ceases, and extending from three
to six inches, as a rule, before it becomes invisible. It
extends slightly further in the case of women than it

does in men.   This aura is very variable, and is greatly
influenced by all the mental and psychic conditions of
the person to whom it belongs.   Its colours vary also
very greatly, but this cannot as a rule be seen through
the screens because they themselves are either dark
red or blue.   It takes a trained clairvoyant to see all
the subtle gradations and variations of colour in the
aura.

HOW TO TRAIN YOUR PSYCHIC SIGHT

The best way to train yourself to see auras of this
character is, perhaps, the following:

1. In a darkened room study the aspect of a good
horseshoe-magnet, either suspended in the air by a silk
thread or placed on a support, with poles up, and vary
the position of the observation until a faint luminosity is
observed at the poles and along the edges of the magnet.

2. In the light repeat the same process, trying to make
out these lines and the extensions and limitations of
the aura.

It must be understood that this vision can be obtained
artificially only through the action of the will, and by a
proper focusing of the eyes,—the perception of auras
requiring a very different focus from ordinary sight,
and this focusing is very often,—nearly always in fact
—different in each of the two eyes.

The attempted focusing of the sight must, therefore,
be made with each eye separately and then with both
combined.   It may happen that one eye only can be
focused for this special vision, or when both are found
available, if both focuses are not identical, the active use
of both eyes at one time may destroy the psychic sight
of the sensitive eye.

THE AURA IN DAYLIGHT AND DARKNESS

It is important to master the faculty of seeing the magnetic aura in the daylight, because more complete details can thus be eventually obtained than in the dark, and this is the only way to learn how to perceive the human aura.

For the purpose of trying one's vision in broad daylight, take a good horseshoe magnet and hold it perpendicularly in front of you,—either against the background of an open outside light, such as can be obtained from looking out from the inside of a room through an open window, or against a near inside background,—for instance a white or dark wall, according to the nature of the light. Then look at the edge of the magnet with one eye only and gradually approach it or slide it away from you, until you obtain the best focus of vision. Look steadily along the same point, until it dawns on you that a kind of a quivering, narrow band of mist or vapour is flowing from the metal and prevents your sight from freely perceiving the object back of it, producing, in fact, a sort of bending of your visual rays. As soon as you realize the presence on the edge of the magnet, of this current of vaporous mist,—which may be compared to the appearance of the heated air which arises in summertime from hot fields,—the first psychic visual victory has been obtained, and the perception of the other phenomena connected with the aura will only need time, perseverance and practice: and, once the magnet is conquered, one may expect to speedily obtain the sight of the beautiful and intricate currents on the human body.

THE STRUCTURE OF THE AURA

After the aura has been perceived, and its general layers distinguished, the student must turn his attention to its structure and colour variations. The question of colour will be treated in the next chapter, which is devoted entirely to that subject.

As to the structure or composition of the aura: If this be studied carefully, it will be found that it is composed in a great variety of different ways, according to the object or person emitting it. Thus, the aura of flowers is very different to that of magnets or human beings.

THE AURA OF FLOWERS

It is a very interesting study to try and perceive, psychically, the composition of the aura of various flowers. For instance, that of the violet is about one-eighth of an inch in thickness, and composed first of a bright light, then a line of dark blue, shading away into a very light blue, all these following the contour of the edge of the leaf. Above these lines is a scalloped or semi-linear string or border of two rows of little purplish-red figures, diamond shaped, very regularly distributed, so as to form two sets of fourteen little diamonds over the space of each small lobe of the leaf. Then, above these, a wave of dark blue mist in crescent form, shading off into light blue.

This is only a sample reading of one flower. Each flower has its own particular aura (some of them being very complex); but it will serve to show the student how interesting a study this can be made. The study of the aura of plants alone, carefully undertaken, would occupy considerable time.

THE HUMAN AURA AND HOW TO STUDY IT

After you have studied the auras of magnets and plants in this way, you should turn your attention to the auras of living, human beings.

Children may easily be studied, and their auras are exceedingly interesting.

Developed clairvoyants are enabled to see several different auras,—each of them being composed of a number of sub-divisions, and each sub-division having a different structure and colour.

It is a good plan to begin the study of the aura by the aid of the chemical screens, before mentioned, in semi darkness: and then to practise viewing the aura without the screens, and, as the eyes gain sensitiveness, to admit more and more light, until it can be clearly seen in the daylight.

In this way your psychic sight will be gradually and naturally developed.

CHAPTER XI

COLOUR AND ITS INTERPRETATION

In the last chapter we learned that there is a psychic atmosphere or "aura," surrounding each animate object, and particularly human beings, and that both the structure and the colour of this varies greatly. We must now inquire first into the *nature* of these colourings, and secondly try to solve the question "what do they mean?" How are we to interpret these colours, realizing that they are but symbols of something which they merely express? The colour of every individual is doubtless somewhat different, and with the same individual it differs at various times, according to his state of health, the mental and psychical changes, etc. The majority of highly-developed clairvoyants agree, however, with C. W. Leadbeater, that the following colours may be distinguished, and that they signify the existing physical, mental and spiritual conditions, as follows:

THE MEANING OF THE VARIOUS COLOURS

BLACK: indicates hatred and malice; anger and hate thought-forms are like heavy smoke. RED: deep red flashes on black ground show anger: lurid red indicates sensuality. BROWN: dull brown-red shows avarice; dull, hard brown-grey selfishness; GREENISH BROWN: with red or scarlet flashes, denotes jealousy. GREY: heavy-leaden shows deep depression; livid grey shows fear. CRIMSON: indicates love. ORANGE: pride or ambition. YELLOW: shows intellectuality; duller tints show it is used for selfish purposes. GREEN: deep blue green shows good qualities, deep sympathy, while grey-green shows deceit and cunning. BLUE: dark, indicates religious feeling; light blue shows devotion to a noble spiritual ideal. WHITE or near white, shows high spirituality. DULL BROWN and blue show selfish religious feeling; dull yellow, low type intellect; APRICOT shows pride; BRICK RED indicates selfish affection and avarice. LIGHT and BRIGHT RED show pure affection. GREYISH-GREEN, with reddish tinge, shows deceit. The health aura is clearly visible to the clairvoyant as a mass of faintly luminous violet grey mist, interpenetrating the denser part of the physical body and extending very slightly beyond.

"It is easy to understand how almost infinite may be the combinations and modifications of all these hues, so that the most delicate gradation of character, or the most evanescent of mingled feelings may be expressed with the greatest accuracy. Many of the colours are unknown to our physical faculties, so that it is impossible to picture them with psychic hues."

When a clairvoyant sees these colours in the aura or surroundings of an individual, therefore, he may feel that the characteristics indicated are present; and the same thing is true when they are seen in the "surroundings" of a returning spirit.

THE SEVEN AURAS SURROUNDING MAN

When the advanced student carefully studies the human aura, he will find that there are a number of straight, white, very fine lines emanating from the body, and particularly the head, which resemble rays of white light. These are magnetic rays which do not deal directly with the psychical condition of the subject.

The innermost aura will be found to consist of five different coloured bands. The first will be pure white, the second light blue, the third darker blue, the fourth lemon yellow, the fifth dark red.

The second layer of the aura will be found to be bluish violet, merging into rose. Those two inter-penetrate one another, forming very beautiful combinations.

The third layer consists of three cloudy zones, the first pink, the second violet, and the third orange.

The fourth layer consists of green, cloud-like waves, tinged with yellow, resembling the golden edges of clouds, behind which the sun is shining.

The fifth will be seen to be slate or indigo in colour, with silver edges.

The sixth will consist of a beautiful light blue, with a whitish golden fringe.

The seventh or outermost aura will be seen to be a greyish mist, of a light violet tint.

The outer aura completes the auric emanation of man, and is the outer shell, as it were, constituting the so-called "auric egg," surrounding every human being,—of which more will be said later.

CHANGES IN THE COLOURS

When a golden yellow light is seen about the head, it may be assumed that such an individual has great intellectuality, combined with spirituality; and in some cases it has been said that this contains a cloud of "gold dust," each speck revolving spirally on itself.

These colours of the various auras are not unchanging, as before said, so that the student must not expect to see them exactly as described. They are greatly modified by the mental and psychical condition of the subject; and the thoughts and emotions which the latter may be experiencing at the time of observation will also affect the aura.

THE COLOURS OF THOUGHTS

Thus fear gives rise to circles of bright rings, spread out in the form of a cone, of varying shades of grey, pink and purple.

A beautiful devotional thought may be expressed in the form of a star of bluish mist, tinged with yellow.

Pity may be seen as a reddish violet cloud, from which issue pointed cones of a brighter pink.

Deception will give rise to a steel-blue mist, tinged with pink and taking the shape of any regular spiral.

Fear may give rise to balls of grey, pink and yellow mist; while fear combined with anger will give forth a blackish grey mist, from which red electrical flashes appear to issue.

THE "AURIC EGG"

These colours extend over the whole of the auric egg, and may be seen by the clairvoyant to be influencing it throughout.

This auric egg which is formed round an individual by the atmosphere or aura, emanating from him, extends both above and below his body, as well as sideways, and is from nine to ten feet in height, and five feet in diameter. If the colours of this auric egg be examined by a clairvoyant, characteristics of the individual may be clearly defined after the necessary practice and development, and the general character of the subject may in this way be discovered and interpreted.

THE COLOURS OF VARIOUS INDIVIDUALS

Of course, it is necessary to make a long study of the aura and to attain a good deal of psychic development before all the details contained in this chapter can be discovered by the student in actual practice. Assuming, however, that you have progressed sufficiently in your studies to be enabled to see the aura of any individual, you may proceed to examine the whole auric egg, with its varied colours. If you do this, you will find them to be about as follows:

A highly developed individual will have a haze of golden light, issuing from the head and extending almost to the top of the egg. Above this will be a faint purplish light. On either side of this golden aura, and issuing outwards and upwards from the shoulders, will be a bluish light, which merges into pink, as it descends to the breast. From that point to the thighs a pinkish light may be seen,—light in some cases, darkish-red in others. About the knees this pink shades off into a delicate green, and this green covers the feet and extends downward almost to the lower margin of the auric egg, where it becomes a darkish blue.

With less highly-developed individuals these colours will vary, according to the tables above given. In some cases the auric egg will be composed almost entirely of greys, greens and browns!

THE PRIMARY COLOURS AND THEIR MEANINGS

In general it may be said that yellow and any bright clear colours, when seen in the aura of an individual, denote strong vitality and active intellectuality. Lilac, blue and violet have to do with spiritual characteristics.

They are associated with simple, unselfish natures, and with those having spiritual aspirations.

Red is directly connected with passions, and particularly anger.

Blue is associated with religious feeling, though if too muddy it denotes selfishness.

The brighter and clearer the colours the better, and they should be as clear-cut as possible.

COLOURS SHOWN BY SPIRITS OR SEEN CLAIRVOYANTLY

These colours are not always associated with the human aura or with any human form. They are often seen by psychics who are developing themselves as cloud-like masses or shapes, which form more or less distinctly in front of them, and appear to take the outline of flowers, flashes, etc. Colours are occasionally seen in dreams, but the dream-images of most people are colourless, or only light grey, like a shadow. They are in fact, "such stuff as dreams are made of."

In general: these colours may be interpreted according to the laws given above, but the precise interpretation of these symbolic messages is a more difficult question to settle. A spirit, who may be trying to communicate and to give a certain message to a medium, may apply this same method of colour symbolism to convey his meaning, but this is often confusing to the psychic and difficult to interpret. We shall come to this question, however, in the next chapter, which is devoted to "Symbolism."

CHAPTER XII

SYMBOLISM is one of the most important and at the same time one of the least understood subjects in the whole realm of psychics and spiritualism. A proper understanding of what it means, and the adequate interpretation of symbols, as presented to the psychic, would prove of great value to every student, and to all those who are undertaking their own psychic development.

WHAT SYMBOLISM MEANS

First of all, it is necessary that you should understand exactly what symbolism means. A "symbol" is a sign for something else which it expresses in a more or less partial and incomplete manner. Usually, a symbol is a sign which appeals to one of the five senses, but denotes not a sensual thing, but the thought lying behind it. Thus the printed word on the page is the symbol of the author's thought, expressed in that word. The poem is the expression of the poet's mind and spirit, as set forth in the words and metre of the poem, etc. Thus, symbols are always only partial and incomplete and represent but a small fraction of the thing they stand for, and we should always be in error if we tried to reconstruct the *whole* of the thing symbolized from what we perceive, by means of our senses.

Before we proceed to the subject of symbolism as studied in psychics and in the phenomena of spiritualism, one other point should be explained: We never see an object in the physical world as it really is!

SYMBOLIZED OBJECTS

We only perceive or realize, through our five senses, various aspects or qualities of the object. Thus, if you are looking at an orange, your sense of *sight* gives you the impression of a reddish-yellow sphere, rather irregular on its surface. Your sense of *touch* tells you that this thing is round, that it is somewhat rough and cool. Your sense of *smell* supplies you with the information that it has a pleasant odour unlike anything else, which is confirmed by your *taste*. In this particular instance the sense of *hearing* does not enter into the question, as it would in many other instances.

Now all these things which appeal to our senses— colour, odour, texture, etc.—are "qualities" of the orange, and not the orange itself. The orange is always something different from all of these, above and beyond them, and is more inclusive than any of these qualities and symbols.

Thus, suppose you took away one of those symbols,— its colour,—the orange would immediately become invisible to you; yet it would continue to exist, though we could never know of its existence.

This shows us clearly that symbols are very inadequate and imperfect representations of a vaster "something" lying behind them, and they represent only a small fraction of the totality of the thing as it really exists.

SYMBOLISM OF SPIRIT

As applied to the spirit of man, we must begin by admitting the rather startling fact that no man has ever

seen it,—no man has ever seen another! All he has
ever seen are the outward features, the form, the facial
expressions of the other; and when our spirits hold
communication with one another in this world, they do
so by written symbols, by motions of the hands or head,
or by means of air-waves, passing from the throat of
one to the ear of the other—all but expressions or
symbols, which are interpreted by us according to a
certain pre-arranged code.

If we did not have this pre-arranged code, it would
be impossible for two intelligent beings to converse one
with another—as may readily be seen when a Chinaman
and an American meet for the first time, neither of
them speaking the language of the other. They try
as best they can to make each other understand what
they are thinking about, what thoughts are in their
minds, but they succeed very imperfectly, or not at
all! The symbols employed are too inadequate to express
their thoughts.

SYMBOLISM IN SPIRIT-COMMUNICATION

Now, all these difficulties we encounter when a spirit
endeavours to communicate with us through a medium,
or directly. It can express itself, as a rule, only very
imperfectly (as will be explained in a later chapter),
and must resort largely to symbols to convey its meaning. Hence we should be very mistaken if we were to
interpret this symbolism literally, or to assume that it
represented the *whole* of the subject-matter which the
spirit desired to convey.

As I said in my book "The Problems of Psychical Research": "Our dreams, as we know, are largely symbolic—the work of Freud and others having proved this

beyond all doubt. It is highly probable that the ravings of delirium are also of this nature, though no one, so far as I know, has yet devoted to their study the attention they deserve. Certainly it is true in mediumistic phenomena, for in trance conditions a large number of the messages, tests and visions seen, are of this nature and character—the symbolism being often so elaborate that the original thought is not perceived. Why this symbolism? The probable answer to this question is, that the message cannot be given *directly*, and that this symbolic method of presentation must be resorted to, in order to get the message "through" at all. There is good evidence to show that a pictorial method is resorted to very largely by the spirits—mediums seeing what they describe, very often, when the more direct auditory method is not resorted to. The spirit presents somehow to the mind of the medium a picture, which is described and often interpreted by the medium. Often this interpretation is quite erroneous, resembling a defective analysis of a dream. Because of this, the message is not recognized, yet the source of the message may have been perfectly veridical (truth-telling).

EXAMPLES OF SYMBOLISM

"Let me illustrate this more fully. Suppose you desire to tell a Chinaman, who speaks not a word of English, to fetch a certain object from the next room. It would be useless for you to say the word 'watch,' because he would not know what the word meant. Probably you would tap your waistcoat pocket, pretend to take out a watch, wind it, look at the hands, etc., in your endeavour to convey to him your meaning. If this were not recognized, for any reason, you would have the utmost difficulty in conveying your meaning to him, and equal difficulty in telling him to fetch the watch from the next room.

"Now, supposing these antics, or somewhat similar ones, were resorted to by a spirit in his attempt to convey the word 'watch'—perhaps to remind the sitter of a particular watch he used to wear. The spirit might well proceed as follows: 'He taps his stomach and looks at a spot over his left side. He seems to wish to convey the impression that he suffered much from bowel trouble, perhaps a cancer on the left side. Yes, he seems to be taking something away from his body; evidently they removed some growth, and he wishes to convey the idea that something was taken from him. Now he is examining his hand; he is looking intently; now he is doing something with his fingers, I can't see what it is, a little movement. Was he connected with machinery in life? Now, he is pointing to the door, etc.' "

THE INTERPRETATION OF SYMBOLS

Such an interpretation of the facts, it will be observed, while describing his actions, is wholly misleading as to its interpretation; the symbolism has been wholly misconstrued; and, inasmuch as the subject probably never died of cancer, had no bowel trouble, underwent no operation, and was never connected with machinery, it is highly probable that the message would be put down wholly to the medium's subconscious imagination, or even to guessing or conscious fraud! Yet, it will be observed, the message was in its inception wholly veridical the fault lying in the symbolism, misinterpreted by the medium.

There is evidence to show that other forms of symbolism are adopted also,—applying to the auditory as well as to the visual presentation of images. It is well known that names are very difficult to obtain by mediums, and this is probably due to the fact that names are not pictures, or visual symbols, and in themselves mean nothing, as a rule. They are merely a combination of letters, having a certain sound.

THE FORMS OF SYMBOLISM

It is generally easier for the spirit to impress a partially developed psychic by means of a picture than in any other way, and for this reason names are difficult to get. Still, in many cases, names are obtained by a picture shown. Thus, the name ''Merrifield'' was in one case given to the psychic as a picture of a number of children, happily playing in a green field.

Among other forms of symbolism are the following: A large key may be shown to the psychic. This may not mean a key at all, but a symbol of success,—the key being the means by which the door of prosperity is opened.

Colours are frequently shown and nearly all colours are symbolic of something or other, and have their definite meaning, as we saw in the last chapter. Strange, weird and horrible figures do not necessarily mean anything bad or anything evil; they may be symbolic of something entirely different, and this is frequently seen in dreams which are composed almost entirely of symbols throughout.

Most psychics, when they are developing, see peculiar specks, clouds and forms shaping themselves before them in space. They are naturally at a loss to interpret and explain these images. While there is much latitude of interpretation, always, in symbolism, the following simple suggestions based on traditional teachings, may be found helpful.

CLOUDS AND LIGHTS

Clouds, if white, may be interpreted as signifying happiness and prosperity, either to the psychic, or to one near and dear to them.

If these seem to recede rapidly, and fade away in the distance, a journey is often indicated. If the clouds appear to be advancing toward you, it indicates that news will shortly reach you, good,—if the clouds are white, bad, if they are dark. If red and lurid, ill-fortune is upon your horizon, for which you must be on constant guard. Black clouds symbolize troubles of the heart.

Tiny moving *specks of light*, if they truly result from psychic development—are said to indicate that you are progressing favourably in your psychic sensitiveness. If these specks are *dark*, however, evil or harmful influences may be about you—for which you must be on a constant look-out. A *"light within a light"* is said to symbolize the presence of some "spirit," desirous of communicating. Should such a sign appear to you, try at once to enter into communication with the spirit-intelligence by asking questions and note whether the light you see endeavours to reply to you by means of some simple code.

Reptiles, and other unpleasant signs, usually symbolize the hidden fears of the psychic; they are symbolized in this way—"externalizing" the subconscious fear-thought of the subject. Root out your hidden fear and apprehension; assert your mastery and fearlessness and the unpleasant sign will always disappear.

THE SUBCONSCIOUS AND SYMBOLISM

The subconscious mind has the faculty of describing in symbolic form thoughts, impressions or influences which come to it, either through the senses, or more directly by telepathic or clairvoyant visions or messages, which are said to be given *through* it by the "spirits." The "spirit" may convey a certain message to the subconscious mind of the psychic, and the message may be "externalized" or presented to the ordinary conscious mind in symbolic form, representing, apparently, something entirely different from the original message.

It is in the interpretation of these symbols that much of the true art of mediumship and psychic development will be found to lie—the better the medium the more expert in the interpretation of these symbols.

At present no general rule can be laid down as to the interpretation of the symbols employed, since these will differ very largely in every case, each medium having his own method of interpretation, and his own form of symbolism.

You must learn for yourself, by repeated experience, what the various symbols mean to you, and thus form a "code" or method of interpretation which you can always follow throughout your future development. A close study of symbolism will yield you very important, practical results, as well as being of great interest in itself.

CHAPTER XIII

TELEPATHY

TELEPATHY, mind-reading, thought-reading, thought-transference, are all terms meaning very much the same thing,—namely, the ability to impress the mind of another person with a definite thought or thoughts, without traveling through the usual avenues of sense. The word "telepathy" was coined by Mr. F. W. H. Myers in 1882, and is derived from two Greek words: (*Tele*—at a distance, and *Pathos*—feeling) and means literally "sensing at a distance." From this it has come to mean "thought-reading" in general, as we now understand it.

HOW TELEPATHY OPERATES

How telepathy takes place we cannot as yet say with certainty. Some scientific men, such as Sir William Crookes, are inclined to believe that vibrations in the ether travel from brain to brain, very much like the messages in wireless telegraphy. Others, on the contrary, contend that this explanation is insufficient, and that we have no proof that such brain-waves exist. As Mr. Myers expressed it: "Life has the power of manifesting itself to life," and this is as far as we can go as yet, by way of scientific explanation of the facts.

It is almost certainly true that telepathy takes place not between the conscious minds of two individuals, but by way of the subconscious, that vast field which we described in Chapter IV (The Subconscious), so that if a message is sent from one conscious mind to another,

it would travel in rather a round-about fashion as follows: From the conscious to the subconscious mind
of A; from that to the subconscious mind of B, and
from the subconscious mind to the conscious mind of
B. In B, the process by which it was conveyed from
the subconscious to the ordinary mind would be that
of "externalization," so frequently seen in dreams, crystal-gazing and other phenomena.

This fully explains to us why it is that we frequently
receive telepathic messages at the moment we are falling
to sleep,—or at least appear to do so. We may have
received the message an hour or two before this, but
its externalization was impossible until the ordinary
consciousness ceased to be so active with the affairs of
the day; and then the subconscious mind had a chance
to deliver its message,—received some time before from
some distant mind.

THE VARIOUS KINDS OF TELEPATHIC MESSAGES

Telepathic messages may be "visual," in which case
they take the form of pictures, figures, written or printed
words, etc.

They may be "auditory," in which case they take
the form of spoken words.

They may be "emotional," in which case the subject
may feel a peculiar depression or excitement.

They may be "volitional," in which case the subject
is seized with the imperative desire to perform a certain
action, etc.

Telepathic messages may originate either in the living
or in the dead. As they are transmitted from the subconscious mind (perhaps under the supervision and direction of the conscious mind), they are often trans-

mitted most effectually during sleep, trance, under the influence of some drug, in delirium, at the moment of death, etc. These messages are most easily received at such times, when the conscious mind is asleep, or in abeyance, and for this reason we have so-called "visions of the dying," ecstasy, trance-speaking and revelation, etc.

TELEPATHY FROM SPIRITS

It is probably true that "spirits" converse with one another directly by means of telepathy, though they understand fully the thought of the other as though the sentence had been fully spoken. Swedenborg tells us that this is the case, and that the telepathic thoughts sent out by a spirit appear to other spirits or to mediums in trance, as clear and "sonorous" as spoken words. If spirits in the flesh can converse at times with one another by means of telepathy, and if disembodied spirits converse with one another by this means, it is only natural to suppose that this is frequently the method of communication resorted to between embodied and disembodied spirits, and all trance-mediums know that this is, in fact, the case. (The larger meanings and applications of telepathy will be discussed more fully in the chapter devoted to Prayer, etc.)

PRACTICAL EXPERIMENTS IN TELEPATHY

The following practical exercises will enable you to prove to your own satisfaction that telepathy exists, and that it can be reduced to a more or less simple process by continued practice.

Select a friend with whom you are in sympathy, physically, mentally and morally. One of you must be the sender or "transmitter," the other the receiver or "recipient." Let us suppose for a moment that you are the transmitter. The recipient should be seated in a comfortable chair at one end of a fairly large room, which must be freely ventilated. It is best that, at least during the early experiments, he should be blindfolded, or that he close his eyes, and sit with his back to you, pencil in hand and pad on knee. He should sit in a semi-darkened part of the room.

For your part, you should sit at a table, facing him (that is, his back) with a pad of paper and pencil before you, have a bright light thrown on the pad of paper, leaving the rest of the room in semi-darkness. Now draw upon the paper a symbol, perhaps a geometrical figure, such as a triangle, circle, square, etc. Look at this figure intently and endeavour to impress it on the recipient. You should not make each trial exceed one minute in length.

HOW TO INSURE SUCCESS

The attitude of mind which you hold during these experiments is very important. You should *will* that your recipient should see the picture presented to him, yet you should not strain yourself in the attempt, and wrinkling the brows, tensing the muscles, etc., will not add to the certainty with which your picture is conveyed, rather the reverse. On the other hand, you should have complete *confidence* in the fact that he will get the impression you are sending him. Never allow yourself for a moment to believe that you will fail. Say to yourself that he has already succeeded in receiving it. Do not allow yourself to become flustered, or worried or anxious. Imagine your thoughts travelling to him in a definite form, either in the *shape* of the object itself, or in the *word,* square, circle, etc., though in that case you must be careful that you do not unconsciously whisper the word so that he hears it!

The recipient, on his part, should make his mind as blank as possible, and note down any pictures or impressions that come to him, no matter how "wild" they may appear. Above all, you must not be discouraged by early non-successes, for these you must expect.

MORE COMPLICATED EXPERIMENTS

After you have succeeded with the diagrams, you may try more complicated pictures, such as playing cards, which are very good for this purpose, as the deck may be shuffled between each draw, and it is easy to calculate the percentage of guesses, since chance would always be 51 to 1 against the subject hitting upon the correct card by accident.

After these experiments you may try some in the transference of pain. Prick yourself lightly in various parts of the body with a sharp needle, or pinch yourself, and see whether the subject can locate the pain correctly on himself. If he is a good subject, he will do so in very many instances, as though the pain were transferred directly to him, and you were pricking or pinching him.

Next you may try a number of experiments in smell and taste. Procure a number of substances such as cloves, nutmeg, pepper, sugar, etc., and smell or taste these in turn, being careful that you are far enough removed from your subject to prevent him from smelling these in the usual way. Many good subjects can tell immediately the substance you are putting into your mouth, the instant you have placed it there. After you have succeeded thus far, you should try to increase the distance between you, until you can perform the same feats, though miles apart.

These simple experiments will prove to you, and to the sceptic, the existence of telepathy. They will render you more sensitive to the reception of messages from distant living minds, and also messages from the discarnate. In this way you will cultivate your sensitiveness to messages of this character, and this will be beneficial to you, provided that you do not carry these practices too far and cultivate your sensitiveness unduly in wrong directions.

Under normal, healthy conditions your mind will not be affected by impressions of this character, since it will be most difficult for you to receive them, as a rule, no matter how hard you may try. The mind always protects itself against too easy access by outside minds. It is very rare indeed that subjects are impressed against their will. Some persons, it is true, believe that others at a distance are influencing them in this manner, and impressing them to do certain things. Many believe that they are hypnotized, etc. But in nearly all cases these beliefs are illusory; they have no foundation in fact. When examined, they are found to rest wholly in the imagination of the subject, and they are frequently but the indications of an unbalanced mind. This does not mean that such persons are necessarily insane, but they would become so were they to dwell upon their imagined grievance long enough, and believe in it after they have been shown repeatedly that such was not the case. It is this persistent will-to-believe in a thing which is not true, that is one of the causes of insanity.

The student who practises telepathy within reason and who has followed the instructions contained in the early chapters of this book as regards fortifying and protecting his own inner nature, need have no fear that telepathic influences or impressions from others will ever affect him against his own will. In nearly all cases, these so-called influences are imaginary; and even should they exist, the subject, who has mastered himself, and who has strengthened his soul from within, is capable of overcoming and repulsing any outside forces of this character, and of preventing any telepathic influence from reaching him, no matter whether this comes from the living or from the dead.

CHAPTER XIV

CLAIRVOYANCE

CLAIRVOYANCE is derived from two French words and means literally "clear seeing."

It means far more, however, in the language of Spiritualism and psychics, and is now used to cover and classify, if not to explain, a large number of different phenomena which some day will probably be explained in other ways. There are various types and kinds of clairvoyance, different authorities having given somewhat different definitions of the various sub-divisions. Thus, the Manual of the National Spiritualists' Association subdivides and defines the various types of clairvoyance as follows·

SUBJECTIVE CLAIRVOYANCE

1. Subjective Clairvoyance is that psychic condition of a human being (who thereby becomes a medium) which enables spirit intelligences, through the manipulation of the nerve centres of sight, to impress or photograph upon the brain of the medium pictures and images which are seen as visions by the medium without the aid of the physical eye. These pictures and images may be of things, spiritual or material, past or present, remote or near, hidden or uncovered, or they may have their existence simply in the conception or imagination of the medium communicating them.

OBJECTIVE CLAIRVOYANCE

2. Objective Clairvoyance is that psychic power or function of seeing objectively spiritual beings, objects

and things by and through the spiritual sensorium which
pervades the physical mechanism of vision, without which
objective clairvoyance would be impossible. But few
persons are born with this power, in some it is developed
and in others it has but a casual quickening. Its extent
is governed by the rate of vibration under which it
operates; thus one clairvoyant may see objectively spiritual things which to another may be invisible, because
of the degree of difference in the intensity of the power.

X-RAY CLAIRVOYANCE

3. X-ray Clairvoyance is a form of clairvoyance which
partakes of the characteristics of the x-ray and seems
to be objective. The clairvoyant who possesses this
power is able to see physical objects through intervening
physical matter, can perceive the internal parts of the
human body, diagnose disease and observe the operations of healing and decay.

CATALEPTIC CLAIRVOYANCE

4. Cataleptic Clairvoyance occurs when the body is
in a trance state, resembling sleep, induced by hypnotic
power, exercised by an incarnate or decarnate spirit, or
it may be self-induced; when in this state the spirit
leaves the body and is able, at its own will or the suggestion of the hypnotist, to travel to remote places and
to see clearly what is transpiring in the places it visits
and to observe spiritual as well as material things in
its environment. While in this state it sometimes happens that the thoughts of the spirit in its travels are
expressed by the lips of the physical body and that
thought-images are conveyed to it through the physical
body. This is due to the fact that there is a spirit

"cord" which connects the body and the spirit and transmits vibrations between them. As long as this cord is not severed the spirit can return to the body, but should it be severed, then what we call "death" would at once ensue. Under this form of clairvoyance there is an interblending of subjective and objective spiritual sight.

TRANCE-CONTROL CLAIRVOYANCE

5. Trance-control Clairvoyance is that psychic state under which the control of the physical body of the medium is assumed by a spirit intelligence, and the consciousness of the medium is, for the time being, dethroned. In this case the controlling spirit is really the clairvoyant, and simply uses the medium's body as a means of communicating what the spirit sees, and therefore the question of subjective and objective spiritual sight, in so far as the medium is concerned, cannot be raised. To some persons who go to mediums for readings and who may become witnesses in trials at law it may not be known that under trance control the medium is to all intents and purposes absent, therefore in dealing with definitions of clairvoyance to be used for the enlightenment of thinking people, judges and juries, it seems necessary for the protection of such mediums to explain what is here termed "Trance-Control Clairvoyance."

TELEPATHIC CLAIRVOYANCE, ETC.

6. Telepathic Clairvoyance is the subjective perception in picture form of thought, transmitted from a distance.

The type of clairvoyance illustrated in class 4 is frequently called "Travelling Clairvoyance" because the spirit appears to travel, after leaving its body, and visit distant scenes. According to the above definitions, this type of clairvoyance is classified under definition No. 1, but other authorities would give it a separate class by itself. When the psychic's mind seems to travel backward along the stream of time, and remembers events which were beyond its normal recollection, we have cases of so called "retrocognition." When, on the other hand, the psychic's mind seems to travel forward into the future and sees scenes and events which, of course, he was otherwise unable to foretell, we have cases of prevision, prophecy and precognition. This latter subject will be dealt with more fully in the chapter devoted to "Prophecy *versus* Fortune-Telling."

We also have spontaneous and experimental clairvoyance, these definitions explaining themselves.

Then, direct and indirect clairvoyance: Direct when no other mind or agency is involved but the psychic's own; indirect when it goes through a roundabout channel and involves some other mind, incarnate or discarnate.

As opposed to "telepathic clairvoyance" we have so-called "Independent Clairvoyance." Also there are cases of "Reciprocal Clairvoyance" in which two persons see one another at the same time and, as it were, exchange their clairvoyant visions.

The type of clairvoyance which is characterized by leaving the body and visiting the spiritual spheres (afterwards returning to reanimate the body) is called "ecstasy."

Mr. Leadbeater divides Clairvoyance into three subdivisions: clairvoyance in time, clairvoyance in space, and direct clairvoyance, in which the astral or spiritual senses are opened up so as to perceive planes of activity now about us.

Clairvoyance may also occur in dreams, crystal visions, etc., and clairaudience (which corresponds to clairvoyance, save that the information is obtained by means of the ear rather than the eye) may be obtained through shell-hearing. In this case the messages are heard rather than seen. (Another technical name for clairvoyance which is sometimes used is "Telæsthesia.")

Clairvoyance may manifest itself in a variety of forms as the above definitions would signify. The most common form is:

Spontaneous Clairvoyance, in which the psychic sees pictures of absent persons and scenes.

Clairvoyant dreams are fairly common.

CLAIRVOYANT DIAGNOSIS

There is a type of mental clairvoyance which enables the subject to see, as it were, into the body of another, diagnosing his disease as though he perceived clearly the conditions present in that individual's body at the time. This form of psychic vision was possessed by Andrew Jackson Davis in a remarkable degree, but is possessed by many psychics of our own day.

Another form of clairvoyance is that in which underground metals and waters may sometimes be perceived. Usually in this case the psychic walks over the ground to be explored with a forked twig or stick in his hands. Suddenly this bends or dips, and where this happens water or metal is to be found. This is technically known as "dowsing" and has been proved to exist as a scientific

fact by Sir William F. Barrett, Professor of Physics, in
the University of Dublin.

Of late years a new and peculiar type of clairvoyance
has been developed by Mr. Vincent N. Turvey of London, a friend of Mr. W. T. Stead. He termed this
"Phone-voyance," for the reason that he receives his
impressions and intuitions, etc., when he is conversing
with a distant friend over the telephone, and then only.
This is a form of sensitiveness which could be obtained
by many psychics if they developed it.

THE EXPLANATION OF CLAIRVOYANCE

Clairvoyance has been explained in a variety of ways.
We may briefly summarize these theories as follows:

1. *The Astral or Spiritual-Sense-Theory.* This may
be stated as follows: Corresponding to each physical
sense-organ (the eye, ear, etc.) there is a corresponding
spiritual or astral sense-organ. We see physical objects
by means of the physical eye and hear them by means
of the physical ear. When we see clairvoyantly, on the
contrary, we see by means of our spiritual eye and when
we hear clairaudiently we hear by means of the spiritual
or astral ear. These spiritual sense-organs function in
a spiritual world (and of course serve the spiritual
body when we die as our physical organs now serve us)
and operate on the spirit plane of activity. If their
use is not cultivated in precisely the right direction,
it may lead to difficulties, for the reason that both the
astral and spiritual sight may be used at the same time,
they may become mixed up, and you may see two worlds
at once instead of one, so that you cannot be sure when
you go outside the door whether you are going to step

on the pavement or into a great ditch, both of which you
see equally clearly before you. Many persons have got
into this condition which takes some time to outgrow.

2. "The Spiritual Influence Theory." According to
this theory clairvoyance is accomplished not by means
of the subjects' own unaided powers, but always through
the instrumentality of a spirit who sees the distant
scenes, etc., and impresses them upon the subjects' mind
telepathically.

TUBES, THOUGHT-FORMS AND DIRECT PERCEPTION

3. "The Astral Tube Theory." According to this
view, the clairvoyant constructs a sort of telescope or
tube for himself out of "astral matter," and through
this he looks. The figures, in this case, always appear
small, and far-off.

4. "By the Creation of a Thought Form." On this
theory, we create a thought form for ourselves in the
locality we desire to visit, and utilize it for the purpose
of observation; we look out of its eyes, hear with its
ears, etc.

5. "The Direct-Perception Theory." This theory
says that outside influences play no part in the phenomena, but that we perceive distant scenes ourselves,
by means of some process of self-projection. But here
we are met with the difficulty that if the psychic is
absent, viewing the distant scene, how can he also be
present in the room, animating his own body and speaking through it as he undoubtedly does in many cases?

Most psychics, when they begin their development,
see shapes and figures more frequently than they experience any other phenomena. They wonder why this

should be. Why should nearly all of us see? (Now
and then, it is true, we come across one who hears more
easily than sees, but he is the exception, not the rule.)

WHY AND HOW WE "SEE" IN CLAIRVOYANCE

The explanation of this fact is probably the following: We use our eyes more than we do any other one
of our senses. We feel that our active consciousness
is more connected with sight than with anything else.
The sight-centres in the brain are more used than any
of the others, and this fact is proved by dreams, in
which we see figures but very seldom hear spoken words.
Again our memory consists mostly of visual symbols. If
we think of a person we call up his image before us,
this being a "memory image." Now, as these parts of
the mind and brain are so active, it is only an extension of this faculty of inducing memory-images, which
enables us to see objects and figures in clairvoyance.
We only have to force this faculty of the mind a little
more than usual to carry it beyond the limitations of
physical sense; whereas, with the other senses, much less
used, we have to do a great deal more of this cultivating
or forcing-process, in order to develop the corresponding
spiritual organs. Clairvoyance and similar faculties depend in many cases upon the partial liberation or freeing of the spirit from the body, and the stimulation of
the corresponding psychic sense-organs into a higher
degree of activity, and so permitting their use. The
following are a few exercises which will be found helpful
in developing this faculty of clairvoyance, according to
our methods of development.

DEVELOPING EXERCISES

1. Seat yourself in a comfortable chair in a semi-darkened room. Mentally construct (i.e., imagine) before you a tube, open at both ends. One end of this tube fits over your eyes, and the other end extends indefinitely outward into space. Imagine that this tube is hollow and that you can see through it perfectly. Turn this tube in the direction of the house of a friend of yours; mentally go into a room and see if you can discover in it any one present,—and if so, who he is and what he looks like. Note what you see carefully. You will be able to verify the next day how far your vision is correct.

2. Construct the tube as before. At the other end of this tube, which you should imagine about one hundred yards long, you must endeavour to see clairvoyantly the face of a friend. Try to distinguish the features of this face, making them clearer and clearer. When you have done this, gradually pull it toward you by an effort of will, until it is only about two or three feet distant. It should then be perfectly clear and every feature distinguishable. When you have succeeded in visualizing this face so clearly that you see it as distinctly as you would if that individual stood before you in life, your progress as a clairvoyant will have made great advances and you may then begin experiments in *influencing* this person at a distance, while seeing his face before you, as explained. *Will* that he should do a certain thing, to think of you at a certain time or see your face float before him as he is busy with his daily occupations. If you practice this persistently, you will

ultimately achieve success, being able to influence persons without doubt.

"POLARIZATION" AND HOW TO USE IT

This ability to influence a distant person or object by means of your will, when directed toward him, has been termed a "polarization," because you polarize a path or channel through the astral atmosphere toward the desired point, and this channel facilitates psychic communication in both directions. A great deal depends, during these experiments, upon your ability to hold the object clearly in your mind's eye and to concentrate upon it. If you do not do this, your efforts will be lost, since you will find there are a great many astral currents, playing to and fro, which tend to disintegrate your own currents set up by you, and unless these are strong you will not succeed in overcoming the astral "cross-currents."

In conducting these experiments you must be *sure*, especially at first, always to *keep* your *consciousness* centred in your own *body*, and not to let it go outward into space along with your thought; your *will* alone must travel outward; you must keep your consciousness within your own physical body. If you do not do this you will be apt to get into trouble. Your starting-point, your "focal centre," as it is called, must always be maintained.

In developing clairvoyance you should remember that faith and belief tend to open up your latent powers and faculties, while disbelief has the contrary effect of closing them and shutting off all further development. This is true in all lines of psychic unfoldment.

Clairvoyance is a faculty possessed by the whole human race in varying degrees, and there are indications that with each generation, its power is becoming greater and greater, so that the time will doubtless come when every one will see clairvoyantly just as we now see with our ordinary eyes. In fact, the possession of strong intuitions and sentiments, sensing the feelings and emotions of others, etc., are but undeveloped clairvoyant flashes, giving you an insight into the mind of the person with whom you are conversing.

FACTORS IN THE DEVELOPMENT OF CLAIRVOYANCE

Concentration is an important factor in the cultivation of clairvoyance. You must train your mind so that you can think of a particular object for several minutes without relaxing or allowing any other thought to enter your consciousness. You must practise gazing at an object until you can do this for two or three minutes without moving your eyes and without fatigue. You should cultivate deep-breathing exercises and, during inspiration, think that you are drawing on the vital energy of the universe, while with each breath you exhale you are throwing off any adverse influences which may have come to you.

"Visualizing" is an important factor in developing clairvoyance. You should get into the habit of calling up before your mind a face you have seen or a scene you have witnessed that day, trying to remember every detail and making it clearer and clearer until you have every detail clear in your mind's eye. You should then endeavour to project it outward into space, as though you were seeing these pictures outside your head as real entities, and not merely as memory pictures.

Crystal-gazing, etc., will greatly help in this.

HOW DISTINGUISH TRUE FROM FALSE

One question which always presents itself to the mind of the student is this: "How can I distinguish the true from the false, real clairvoyant visions from memory pictures and hallucinations?" It is extremely difficult to do this, particularly for the beginner, and this ability to distinguish comes only with prolonged practice and experience. A lady of my acquaintance has all her life been enabled to distinguish between phantasmal figures of the living and those of the dead. That is, she could tell by looking at the figure whether it represented a living or a dead person. Again she was always able to tell whether this was a genuine or helpful intelligence or whether it was an evil or lying one. This ability to distinguish cannot be gained in a day, it must come by practise; and the beginnings of genuine clairvoyance can only be ascertained by experiment and by following up the visions and figures which appear to one, and see whether they lead to anything definite in the way of progress and enlightenment or not.

As to the various symbols and colours which appear to the clairvoyant, these should be interpreted according to the rules laid down in Chapters XI and XII, devoted to these subjects. Clairvoyance in dreams will be discussed in the next chapter.

A few words in conclusion, as to the theory of clairvoyance: It will be remembered that I enumerated before (pp. 117-24) the various theories which have been advanced to explain Clairvoyance, the astral-sense theory, the direct vision theory, etc. Our own view of the matter is that all of the explanations previously proposed are in a sense true, that is they are all true in particular cases, but that none of them explains all types of clairvoyance. The spirit-influence-theory is true in some instances, the direct-clairvoyance-theory is true in others, and the astral-sense-theory is the correct one in still others, etc.

THE MYSTERY SOLVED

As to the difficulty which is presented by the fact that a clairvoyant can animate and speak through his own body while he is psychically active elsewhere, this is explained by assuming that only a portion of his total psychic self remains behind, and that the more active spirit-part performs the journey or "excursion,"—this part being connected to the physical body by means of a cord or connection which unites it to the latter. This connection is sometimes called the "Silver thread," and is the channel of communication between the spirit and its body, while the former has gone on its excursion. If this "thread" were to snap or become broken, death would take place, since the spirit would be unable to return and reanimate the body. Such accidents are extremely rare, but they have been recorded from time to time in the past. In clairvoyance, the connection remains, of course, intact, and communication back and forth, between the body and the absent psychic self, takes place along this cord or thread. In still other types of clairvoyance, on the other hand, no actual excursion or leaving the body takes place; the active consciousness remains in the body, animating it, while the clairvoyant vision takes place through the psychic "telescope" or "tube" before mentioned. In such cases, there is no difficulty in accounting for all we see, but this is not such an advanced type of clairvoyance as the former, in which the psychic self leaves the body, and goes on trips and excursions of its own.

CHAPTER XV

DREAMS

DREAMS usually take place during sleep, though there is a peculiar form of imaginative picturing which may occur sometimes during the waking hours and which we call "day-dreaming." Of these I shall treat in another place. Speaking first of ordinary dreams, I may begin by pointing out that, before we can understand them, we must know a little of the nature and phenomena of sleep in which they occur. We all spend about one third of our lives in sleep, so that this is a condition which we should certainly know something about if we possibly can!

WHAT IS SLEEP?

Various theories have been advanced in the past to explain sleep, but no satisfactory theory has ever been fully accepted. Thus we have the so-called "chemical theories," which endeavour to account for sleep by assuming that certain poisonous substances are formed in the body during waking hours and are eliminated during sleep. Others have suggested that sleep is due to peculiar conditions of the circulation of blood in the brain; still others that the action of certain glands explains sleep; others that muscular relaxation accounts for it; others that the lack of external stimuli is sufficient to induce profound slumber. All of these theories have been shown insufficient to explain the facts. We shall never arrive at a satisfactory theory of sleep, doubtless, until we admit the presence of a *vital force*

131

and the existence of an individual *human spirit*, which withdraws more or less completely from the body during the hours of sleep, and derives spiritual invigoration and nourishment during its sojourn in the spiritual world. We shall speak of this more fully presently.

THE SEVEN COMMON DREAMS

For the present we must explain, to begin with, the common types of dreams, and show how they are to be accounted for:

There are seven types of dreams which, it is said, everybody experiences at one time or another in his life. These are:

1. The falling dream,
2. The flying dream,
3. The dream of inadequate clothing,
4. The dream of not being able to get away from some beast or injurious person or thing that is pursuing,
5. The dream of being drawn irresistibly to some dangerous place,
6. The dream that some darling wish has been gratified,
7. The dream of being about to go on a journey and being unable to get your things into your trunk, etc.

Some of these dreams are to be explained in one way, some in another; but broadly speaking, it may be said that all ordinary dreams, such as the above, or others of a like type, are due to one of three causes:

1. Physical stimuli,
2. Subconscious mental association,
3. Subconscious imagination.

In addition to this subconscious field there is also another: The *Super-Conscious* of which we shall speak later; but as this is not recognized by "orthodox" psychologists of today, we shall not discuss it for the present.

CAUSES OF DREAMS

Physical stimuli give rise to dreams in this way: The dream of inadequate clothing for example, is doubtless produced by chilling the surface of the body, this, in turn, usually being due to the bed-clothes falling off onto the floor. The dream of falling is probably due to the fact that, by lying too long in one position, the blood-supply is cut off, causing loss of sensation in the under part of the body, this in turn giving rise to the idea that we are not supported, consequently that we are flying or falling, etc. If a book is dropped, this may be symbolized as the report of a pistol in a duel, etc.

Association causes many dreams in the following way: One idea or object of the mind brings up another connected with it, more or less directly, as it would be in life, and the whole storehouse of the subconscious mind is drawn upon in these associations, so that dreams are far more varied than our conscious associations. In addition to this the third factor namely "imagination" is greatly enlarged and given free play, for the reason that the conscious, logical mind is dormant, to a great extent, and hence the wild flights of imagination, which we take in sleep, are possible.

THE INTERPRETATION OF DREAMS

It is because of these facts that nothing appears absurd to us while we are dreaming; no matter how ridiculous a situation may be, it never seems so to us until we are awake and able to reason over it. The curious medley of thoughts composing most dreams, presents a striking resemblance to the ravings of delirium and insanity, and various medical authors have written books, aiming to show the close similarity between dreams and such insane "wanderings."

It has been shown, of late years, that almost all dreams, however illogical they may appear, are in reality more or less consistent, and that a logical strain or undercurrent may be found running through them, if they are analysed and examined carefully enough. The celebrated Dr. Freud of Vienna has worked out an elaborate system of dream-interpretation, based on his exploration of the subconscious mind, and those who may be interested may consult his recent work on "Dreams." He traces most dreams to early childhood impressions, and believes that they express, as a rule, suppressed wishes which have slumbered in the subconscious mind of the dreamer and are "externalized" in this form.

Dreams have, in fact, been compared to the bubbles which break upon the surface of a pond of water. In both cases they have risen upwards through the lower strata and we see the finished product only. What gave rise to this? This is a subject for further investigation.

HOW TO ANALYSE DREAMS

If you wish to ascertain the causes of certain dreams, you may often do so in the following manner:

Place your subject, who wishes his dream analysed, in a comfortable chair, seated in a quiet room in semi-darkness. Set going a metronome or ask him to listen to the tick of a large clock. While doing this certain

images and impressions will arise before his mind. Ask
him to tell you what these are. As soon as he has done
so, question him as to the origin of these, etc., and, by
continued questioning (going deeper all the time into
his mind) you will ultimately find out the origin of
his dream. An example will prove this:

A lady of our acquaintance went into hysterics every
time she smelled plum-pudding! She could not account
for this. One night she had a dream in which she saw
herself cooking pudding in the kitchen, and woke up
in a great state of fear and excitement. Analysis of
this dream showed that when she was a little girl, she
had been left alone in the kitchen while pudding was
cooking, and that the pudding had burnt and nearly set
the house on fire. She saw herself running from the
kitchen and screaming. As soon as this was discovered
she no longer experienced any fear or unpleasant sensations while smelling any kind of pudding; she was, in
fact, completely cured! The subconscious fear had been
removed and its evil effects ceased.

Many similar fears, which terrorize our dreams and
cause nightmare, could be shown to be due to these early
childish impressions, were we to analyse carefully enough
such dreams.

THE SYMBOLISM OF DREAMS

The main characteristic of nearly all dreams is their
symbolism. Of all our experiences, dreams are doubtless the most symbolic. They represent certain wishes,
desires, emotions, thoughts, etc., which fill the subconscious mind, and which become associated together, forming what are known as "complexes." These thoughts,
as they become externalized, are presented in symbolic

form. Thus, a snake may be a symbol of fear and hatred; an angel may be a symbol of love; a key may be a symbol of success, etc.

SEEING AND HEARING IN DREAMS

Few people see colours in dreams. The shadows and figures which make up nearly all dream images are colourless or are of the consistence of light smoke. Just *why* this should be so we do not exactly know; but some artists, who deal a great deal in colour, experience dreams in which all the characters are clothed in gorgeous and highly coloured robes. This, however, is the exception, not the rule.

Again, why do we all *see* figures, scenes, etc., in our dreams? We very seldom *hear* either music or spoken conversation. Words are more rare than pictures. The reason for this is probably that we use our eyes more continually and more consciously than we do our ears, and for this reason the visual images are more easily expressed than the auditory symbols. Smell, taste and touch are even more rare factors in dreams than hearing.

Many persons experience peculiar visions while falling to sleep. They see hundreds of tiny faces before them in the dark, which may condense into one, and this becomes larger and larger and finally vanishes, etc. These are well-understood, and need cause no fear or anxiety.

PERSISTENT DREAM IMAGES

On the other hand, dream-pictures or images may continue for some moments *after* awakening, and these are called "persistent dream images." Thus, Dr. Abercrombie mentions an instance of a medical friend of his, who, having sat up late one evening, fell asleep in his

chair and had a frightful dream in which the prominent
figure was an immense baboon. He awoke with the
fright, got up instantly and walked to a table which
was in the middle of the room. He was then awake
and quite conscious of the articles around him, but
close to the wall of the apartment he distinctly saw the
baboon, making the same grimaces which he had seen
in his dream, and the picture continued visible for
about half a minute! This is a good case of persistent
dream-image.

WHAT IS SOMNAMBULISM?

Occasionally the muscular system becomes active during sleep, instead of the senses only, and then we have
cases of *somnambulism*, in which the patient walks and
talks in his sleep, etc., and even does consecutive mental
work. This shows a too active condition of his subconscious mind, which should be checked by proper treatment. It is extremely dangerous to wake any one suddenly in the middle of an access of somnambulism. If
the patient talks in his sleep it may be very interesting,
at times, to converse with him in a low tone and see
whether or not he will reply intelligently. Many cases
are on record in which valuable information has been
obtained in this way, not only about the subject, but
about distant scenes and even about his spirit friends.

It is possible, also, to cultivate automatic writing
with a good somnambulist, and, in one case known to
us, the patient went to bed with a planchette board tied
to her hand, the pencil resting on a large sheet of paper
and when she awoke in the morning it was covered with
interesting messages! This is an experiment which the
enthusiastic student would do well to repeat.

I must now speak of "superconscious" dreams, in which we are brought into contact with a higher plane of life and activity in the same way that we are in contact with a lower plane during many of our ordinary dreams. When this happens, we experience so-called "super-normal dreams" of which the following are types:

SUPERNORMAL DREAMS

1. The telepathic dream, in which telepathy occurs during sleep between a distant living mind and the sleeping mind of the subject. Information is thereby imparted, which the sleeper could not possibly have known. For instance, in one case, the sleeper's brother appeared to him and notified him of his own recent accident which proved to be true. In another case the sleeper dreamed that a friend of his told him something which also proved to be true, etc. These are so-called telepathic dreams.

2. Clairvoyant dreams in which the subject sees distant scenes and his vision subsequently proves to be true. In such cases, the dreamer apparently leaves his body and travels to the locality in question.

3. Premonitory dreams in which a vision of the future is obtained. In a few days, weeks or months, as the case may be, this dream-vision is fulfilled to the letter.

4. Spirit-Communication during dreams in which a discarnate spirit apparently appears and gives a message to the sleeper, either of consolation, or perhaps tells him an important item of news which he should know. In cases of this character we border very closely upon the medium-trance and true mediumship. In rare cases, communication has apparently been established in this manner.

EXPERIMENTALLY INDUCED DREAMS

It is possible to *induce* telepathic dreams experimentally in another, and you will find it most interesting to endeavour to do this, or to serve as the subject for others who endeavour to induce certain dreams in you during your sleeping hours.

In such cases the sleeper has only to describe as carefully as possible his dreams on awakening. Those who endeavour to impress the dreams upon him must picture in their minds a clearly-formed series of images—allowing these to float before them in space, endeavouring to impress each one in turn upon the sleeper by the power of *will*. After a little practice these experiments will often be found to succeed.

It is possible to control our own dreams, to a certain extent, if we desire to do so. Thus, on falling to sleep, you may will that you experience dreams of a certain character, and if you set about it rightly you can obtain these in many instances. Help is frequently given in this way. I know of several cases in which a subject has fallen to sleep after mentally suggesting to herself that she would receive enlightenment, help and counsel through her dreams concerning the difficult problems of her daily life. In practically every case this was given, though often in somewhat symbolic form. If this were cultivated, it would prove a useful adjunct to our daily lives.

REMEMBERING DREAMS

Another good experiment, which the interested pupil should make, would be to endeavour to catch himself falling asleep, that is, to analyse the gradual loss of consciousness in his own person which occurs as he is falling into sleep. Some people can catch themselves in this way and others cannot. Those who are wide awake one minute and asleep the next will probably never make first-class mediums. Those who linger in the borderland the longest, are those who are naturally most psychic.

Another test is that of remembering dreams. If you can remember clearly a large percentage of your dreams, you are probably quite psychic. On the other hand, if you remember nothing that has occurred during sleep, you are more or less matter-of-fact, and, unless you are the exception, probably will not attain any very great development along psychic lines. It is unwise, however, to cultivate to too great an extent this habit of remembering your dreams. If you do, you will thin the wall which separates your dream-life from your waking-life, and if this becomes "perforated," trouble may result. Keep the two distinct, therefore, after your first initial experiments at introspection and dream analysis.

CHAPTER XVI

AUTOMATIC WRITING

AUTOMATIC writing means writing which is performed without the use of the conscious mind, that is, writing which is performed by the unconscious muscular energies of the hand and arm; hence automatic or non-conscious writing. A pencil is taken in the ordinary way and held over a piece of paper, and in a short time it will be noticed that slight movements of the pencil occur, making scrawling marks on the paper. As time goes on, these marks become more and more consistent and consecutive. They begin to form circles, hooks, etc., until letters, then words, and finally whole sentences are written out.

HOW TO OBTAIN AUTOMATIC WRITING

The best way to obtain automatic writing is to hold the arm clear of the table,—that is, so that neither the wrist, nor the elbows, nor any part of the arm touches it. In this way a certain amount of fatigue is soon induced in the arm, and, as soon as this occurs, automatic writing tends to begin.

In obtaining writing of this character you must be careful to abstract your conscious guidance from the hand as much as possible, leaving it to itself. Do not try and write anything of your own volition; let it guide itself, even if it writes nonsense at first.

Some persons obtain writing more easily if the pencil be placed between the first and second fingers, but what-

ever way is most convenient to you should be adopted
in cultivating automatic writing.

Make the mind as blank as possible. After a time you
may be able to think of other things at the same time,
carry on a train of conversation, read a book, etc., at
the same time that your hand is writing the messages;
but it is improbable that you will be able to do this at
first. The chief thing is to make the mind blank and
await results.

TWO IMPORTANT RULES TO FOLLOW

When developing automatic writing, you should sit
for *not longer* than fifteen or twenty minutes daily and,
if possible, *always at the same time*. It is very important that these two rules be observed, for two reasons:

In the first place your spirit friends, who are, we are
told, trying to help you in your writing, would come to
assist you at certain stated times more easily than irregu-
larly, especially if you told them exactly at what time
to come. It is a good plan to say aloud, just when you
have finished the writing: "Good-bye, tomorrow at the
same time we will sit again for the messages!"

In the second place the time should be limited. When
you obtain writing of this character you are apt to get
so interested in the results, when once messages begin to
come, and so curious in seeing what your hand says,
that you will lose all account of time, and, if you have
nothing urgent to do, are liable to run on hour after
hour writing automatically and replying to the messages
you receive. If you do this for any length of time you
will break down the "wall of defence" which normally
exists against outside influences, and of the importance

of which I have spoken so often in the preceding
chapters. Mr. W. T. Stead, the well-known journalist
and spiritualist, once stated that he considered these
two warnings of the utmost importance, and attributed
his own success (and the fact that he had never en-
countered any difficulties or any trouble in his automatic
writing) to the fact that he had heeded strictly this
advice.

HOW AUTOMATIC WRITING IS ACCOMPLISHED

Automatic writing is doubtless performed by the subconscious muscular action on the part of the hand and
arm of the writer, that is in the majority of cases. But
this does not serve to "explain" it, as many people
believe. Granting that the actual writing is obtained
in this way, the question remains "How about the
information which is often obtained by means of the
writing, information which the writer could not possibly
have known by any normal means?" For instance,
suppose you are sitting at your table, pencil in hand,
waiting to see what is written. Your hand writes: "I
am James Valentine. I was killed in a railroad accident
this afternoon at four o'clock." Granting that your
own hand actually moved the pencil to write this message, where did this piece of information come from?
How did your mind know what to write, and the fact
that James Valentine had been killed? That is the question which remains to be solved, and is the one which
the majority of scientists who have undertaken to investigate and explain these phenomena slur over and
leave altogether unexplained. In many other cases, also,
the power seems to be greater than the medium alone

could have produced, and in such cases an outside power was doubtless employed, as in many "physical phenomena."

THE CHARACTER OF THE MESSAGES RECEIVED

Many of the messages you receive, especially at first, will doubtless prove incoherent and disconnected, like dreams; in fact they *are* dreams, only instead of seeing these thoughts in visions, they are written out by your own hand. In both cases, however, it is your dream-consciousness (subconsciousness) which originates the messages or the visions.

In many cases, however, clear and consistent messages are written and these may be supernormal and show evidence of telepathy, clairvoyance, premonition, or spirit-communication, just as dreams do. Many mediums obtain their messages direct by automatic writing. Mrs. Piper of Boston, in many ways the most famous medium in the history of Psychics, obtained nearly all her communications in this manner. In her case, she passes into very deep trance while writing and has to be supported by cushions. In your own case, it is improbable that you will go into trance at first,—though you may have a tendency to do so, and if you begin to feel sleepy or drowsy during the writing, you should give way to this and allow yourself to pass into the trance-condition. In this state many of the best messages are obtained. It is advisable, however, to do this for the first few times only in the presence of an experienced medium or psychic, who can attend to you during the period of trance, and who will ask questions for your hand to reply to, etc.

PHENOMENA WHICH MAY OCCUR DURING THE WRITING

This feeling of drowsiness appears very often in automatic writing, but it is not universal. Many mediums who obtain remarkable messages in this manner have never passed into trance and have no desire to do so; they remain perfectly normal throughout.

It may be that when you begin to write, your hand and arm will shown signs of insensibility; that is, it will lose its sensation and any feeling of pain, etc. It becomes, as we say, anæsthetic. You may be quite unconscious of this fact and only discover it by an accident. A good plan is to have a friend test it for you. When you are obtaining automatic writing, close your eyes and turn your head away, then ask your friend to prick you very lightly with a needle in various parts of your hands and arms and see whether you experience any pain. It is quite possible that you will not do so, even if the pricks are severe. It is curious to note, however, that these pricks are noticed by the subconscious mind, for it often happens that the hand will write automatically: "You are hurting me," or "you pricked me in the third finger-joint," or something of the kind while you yourself might remain totally ignorant of the fact, so far as any consciousness of it was concerned. It is important for you to remember that automatic messages, like all messages of a like character, must be judged and accepted for what they are worth.

MORE PHENOMENA

Some of these messages are very remarkable, and contain sound advice which can be followed with profit. Some apparently originate from those spirit friends who claim to give them. On the other hand, many of them are foolish, lying or merely silly, so that here, as in all other cases, discrimination must be used, and you must exert your own common-sense and judgment in the matter of accepting these messages, and you must see to what extent you may be willing to abide and profit by the advice given.

It sometimes happens that automatic writing forms letters, but these appear curiously shaped and the words cannot be read; sometimes it begins at the right hand side of the page and writes toward the left, like Hebrew. When this is the case it is always a good plan to hold the sheet of paper up to a mirror to see whether the writing can be read in this way. If so, the writing has been merely reversed, and is what we term "mirror-writing."

Some persons can write with the left hand as well as with the right, but usually this is not the case, except with left-handed persons. The reason seems to be that the left hand is poorly developed as a writing machine. For this reason, we can hardly expect any intelligence who may desire to give messages, to find this an easy way of expressing them! Still it may be tried after writing has been obtained by the right hand.

Occasionally messages are given in foreign languages or in queer tongues, unknown to the sitter. These may be genuine messages and if they come in a language unknown to you, you may be more or less assured that they emanate from some spirit friend who speaks the language in question. Occasionally, however, your hand will write "gibberish," and there are many cases on record where this has been done and no true language has been written.

HOW TO DEVELOP THE POWER

It is a good plan to sit in a semi-darkened room while obtaining automatic writing, in a comfortable position and with the mind as free from care and preoccupation as possible. Automatic writing may, however, be obtained in a light room, if desired.

Telepathic experiments may be tried in this manner: A friend of yours may try to impress upon you certain words, cards, figures, etc., which your hand writes automatically.

The writing, you must remember, is only another method for the subconscious to express itself to the conscious mind, and happens to be a motor-channel rather than a sensory-channel. If it were the latter, you would see or hear the message instead of writing it. In both cases, however, the phenomena represent mere "emergence."

It is not necessary to write automatically with a pencil, for a Planchette-Board, Ouija-Board or some other apparatus may be used for this purpose. Indeed, this is a much simpler method to begin with, and writing is obtained more easily than by the pencil alone. Most people find it more satisfactory, however, to discard these instruments later on, and employ the pencil direct.

The above rules should enable the student to obtain automatic writing in a comparatively short time. *Patience* is required here as elsewhere. Hold the mind in a receptive attitude, send out a mental call for guidance and wisdom, and do not come to the conclusion too quickly that the messages you receive are nonsense. Often a jumble of letters that, at first sight, mean nothing, may form a very significant message, when rightly interpreted.

CHAPTER XVII

CRYSTAL-GAZING AND SHELL-HEARING

CRYSTAL-GAZING means, simply, the practice of looking into a ball of crystal, glass or some similar substance and endeavouring to see within it pictures or images which apparently present themselves to the eye, while thus gazing at it.

Crystal-Gazing is very ancient. The Egyptians used it in their practices of divination, and, throughout ancient history, we find traces of this magical art. In the middle-ages it was revived, especially by the learned Dr. Dee, who lived in the reign of Queen Elizabeth in England, and who employed a seer or "scryer" by the name of Kelly. Dr. Dee wrote a book on his researches, which work is now classical.

In more recent times Crystal-Gazing has been made a subject of special study by the Psychical Research Society and several books may now be had upon the question. It is a very simple and at the same time one of the safest means of psychic development. It is not necessary, as a matter of fact, to employ a crystal or even a glass-ball, particularly if you are a good subject, but it would greatly help matters if you did possess one, and we should advise the student to procure one if possible and use this for purposes of experimentation.

HOW TO BEGIN

The best way to begin is to procure a crystal of at least three inches in diameter, larger if possible, and mounted upon a slender wooden stand. The stand and crystal should be placed against a background of black felt or cloth, and the crystal should be shaded with more cloth of the same character, so that there is no high-light anywhere upon it, that is, no point upon which the sun's rays fall, making it a bright spot. If the outlines of the ball appear a little cloudy and uncertain, owing to the semi-darkness, this will often help matters.

Place yourself in front of the ball, your eyes being about a foot from its surface. You should be seated in a comfortable chair, your eyes shaded from the light and relaxed in body and quiet in mind. Gaze steadily at the crystal for a few minutes; do not strain or focus the eyes particularly upon any part of the ball or try to see into its interior. Do not blink the eyes more than you can help; at the same time do not strain them by trying to keep them open for any length of time without blinking. Do not let your eyes wander from the ball nor your attention relax from the subject in hand. Do not let your eyes stare vacantly, but look intently at the ball without undue strain or concentration. Try not to think of anything in particular during the process of this gazing; make the mind fairly blank, at the same time do not allow yourself to become sleepy or the mind to become totally blank to outside impressions.

It is inadvisable to keep this up for more than five minutes at a time at first, for if you do you will find that your eyes will become strained and will "water" after you leave off the experiment. If this is the case you may be sure you have continued gazing for too long a period. As in automatic writing, it is advisable, if possible, to sit at the same time every day, while develop-

ing and for the same length of time each day. This
time may be lengthened as you progress, though it is
usually found unnecessary to look into the crystal for
more than a few minutes at a time, for you cannot get
consistent, long-drawn-out visions, as you can Automatic-
Writing.

EXPLANATION OF CRYSTAL-GAZING

Crystal-Gazing depends largely upon the ability possessed by the psychic to "visualize" or express in pictorial form, thoughts and images which arise from the
subconscious mind. The majority of crystal visions are
of this character. You must not assume that because
you see figures in the ball that these figures are really
in that place,—that is, that they are objective or external
and exist within the crystal. No,—they are mental
pictures or hallucinations, but they are expressed or
externalized in this way.

For example: You may think of a friend's face and
bring it up vividly before your mind's eye, as a memory
picture. Now, in ordinary life, the process of externalization ends there, but if you are a good visualizer
you can carry it further, and actually project into the
crystal the picture of your thought, placing it *in* the
ball, where you will see your friend's face clearly reflected from within its depths. But your friend is not
really in the ball; it is merely your mental conception
or picture of him. Nearly all crystal visions are of this
character, as before said.

SUPERNORMAL CRYSTAL VISIONS

Crystal visions, however, often contain information
and messages which the sitter could not possibly have

known normally, and which are conveyed to him by this means. For instance, you may look into the ball one day and see, acted before you in the crystal-vision, a tragedy in which some friend of yours plays a part. You know nothing whatever about this, yet later on you receive from this friend a letter, telling you of the details of the tragedy in question. Your vision has proved correct. It is authentic and "supernormal" in character. Thus you will see that crystal-visions are more than mere empty visions or hallucinations. The character and content of these pictures often convey striking information and they may be telepathic, clairvoyant or premonitory,—just as dreams are,—or they may represent genuine spirit messages, conveyed from some deceased friend or relative. It is no unknown thing to see words written in the ball as though you were reading handwriting. A friend of mine once looked into her crystal ball and saw within it a newspaper notice of the death of her dearest friend. She was totally ignorant of the fact and only learned it later on. This same lady who is a writer, has the power of projecting or placing in the crystal, at will, figures or scenes which she conjures up before her and when they are in the ball, they will continue acting out the parts assigned to them, just as they would in a dream,—for the figures seen in the crystal are not inert and motionless, but move about and appear to have life and motion of their own. On many occasions when this lady placed the characters of a novel she was writing into the crystal by an effort of will, she was enabled to see them there, and they frequently enacted certain scenes which gave her a good idea for the continuation of the plot of the story!

In such a case, you will see, crystal-gazing performed a very useful and practical service.

HOW TO DEVELOP THE POWER

You may develop the power of visualizing in yourself, which is extremely important, by such simple imagination-exercises as the following: Ask yourself a question, such as "What was the colour of Mother Hubbard's dog?" "Was Jack, the Giant-killer, dark or fair?" "Was Helen of Troy tall, or small and slender?" Such questions as these should bring up before your mind's eye an immediate answer in the form of a mental picture of the person or event in question, and if they do *not* do so, you may be sure that your power of visualizing is not good and will have to be developed before you can have clear crystal visions. If your power of visualizing is extremely good you will probably be enabled, after a certain length of time, to dispense with the ball altogether, and see your visions upon a white or black background, by concentrating upon it, and finally anywhere in space that you may choose to induce them.

When you have arrived at this stage of development, however, you are very far along the path of successful mediumship!

CLOUDING AND VISUALIZATION

If you are to obtain crystal visions you will probably notice that, just before the vision appears, the ball will cloud over as though a blackish grey mist were filling it, or were interposed between your eyes and it. This "clouding," as it is called, is well known and is a symptom of oncoming visions.

If, after sitting for five minutes every day for a couple

of weeks, you do not obtain any visions at all, you may rest assured that you are a very poor visualizer, and will probably not succeed in this direction.

You might try, however, one simple experiment for a few days longer. Gaze at a bright and highly coloured object upon which the light is falling for about a minute; then close your eyes for a few seconds, and then look at the ball. If you are ever to see anything you should, after a few attempts, see within the ball a duplicate of the object you have been looking at, in its complementary colours.

It is asserted by a certain school of occultists that the visions seen within the crystal are not invariably subjective or hallucinatory, but are real entities, and that the figures have an independent existence apart from the seer. This, however, is a complicated question which is unsuitable for a primary book of instruction upon psychic development such as the present. It will therefore be omitted from consideration with this brief mention.

SHELL HEARING

If you place to your ear two large conch-shells, you will hear a peculiar rushing or roaring sound as of the sea in the far distance. This is only natural and probably due to the air within its cavities and the resounding properties of the various curves of the shell. So far all is simple enough, but many persons, who are slightly psychic, as soon as they place the shells to their ears, hear distinct and characteristic sounds, usually in the form of whispered or spoken words. These words may be inarticulate, they may be incoherent like dreams, they may repeat your own name time after time, or they may convey systematic and definite messages.

As in the case of Crystal-Gazing, Dreams and Automatic-Writing, Shell-Hearing is a method of "externalizing," or expressing in outward form, the thoughts and auditory messages of the subconscious mind. But they may be more than this. They may at times embody telepathic, clairvoyant or premonitory messages, or they may represent genuine spirit communication. It all depends on the content of the message, and upon the character of the word spoken, just as in Planchette-Writing. If you obtain a jumble of nonsense you may be sure that it is the product of your own subconscious activity, but if you obtain a characteristic and direct message, you may have reason to believe it emanates from the friend it purports to proceed from. In Shell-Hearing it is the same.

IMPORTANT WARNINGS AND ADVICE

If the messages are nonsensical they should be disregarded; if on the other hand they are interesting, clear-cut and are proved to be correct, you should regard them as possibly genuine mediumistic messages and they should be judged and valued by you accordingly.

In all cases of this character, here as elsewhere, you must use your own critical judgment and common-sense upon the messages you receive. Shell-Hearing is certainly one of the clearest, at the same time one of the most pleasant methods of receiving communications that can be employed. The voices which you hear may be recognizable or unrecognizable. It is the former that are a good proof of authenticity. They may develop by themselves or emerge from a confused babble of sound. Unrecognized voices will often utter warnings or convey information of this character. Human voices are not always heard in the shell, but occasional musical and other sounds which can not easily be described.

Finally, an important warning should be heeded. If after discontinuing Shell-Hearing, you continue to "hear voices" you should immediately drop all experiments for some days, as this phenomenon of "insistent voices" is one of the first symptoms of danger. As long as the manifestations are well controlled, you may feel that you are on the safe road, and developing as you should; but if they begin to get beyond your control, you should stop Shell-Hearing for some time, until you have strengthened your inner self, to such an extent, that you think it advisable to continue experimenting again in this direction.

CHAPTER XVIII

SPIRITUAL HEALING

SPIRITUAL Healing means that mentally or physically sick persons may be, and are, healed by the power of a spiritual energy, operating through the body of a certain medium, or more or less directly, without his agency. It is distinct from hypnotism, mesmerism, magnetic healing, faith-cure, mind-cure, or any other kind of healing whatever, and must not be confused with them. All these other curative measures depend upon suggestion, or upon the hidden and unknown powers of the human body to effect the cure. But spiritual healing is more direct; it is not the medium who heals in this case, but a form of spiritual energy, which operates through him.

WHAT IS SPIRITUAL HEALING?

Spiritual healing is effected in various ways, as the following definitions, adopted by the National Spiritualists' Association, will show:

"(a) By the spiritual influences working through the body of the medium and thus infusing curative, stimulating and vitalizing fluids and energy into the diseased parts of the patient's body.

"(b) By the spiritual influences illuminating the brain of the healing medium, and thereby intensifying the perception of the medium so that the case, nature and seat of the disease in the patient become known to the medium; and the herb or other remedy which will benefit the patient also becomes known to the medium.

"(c) Through the application of absent treatments, whereby spiritual beings combine their own healing forces with the magnetism and vitalizing energy of the medium, and convey them to the patient who is distant from the medium, and cause them to be absorbed by the system of the patient."

It will be seen that these definitions not only cover the facts of spiritual healing, but also absent treatment and psychic diagnosis.

Advice is given on numerous occasions by the "spirits," as to the exact course of treatment to be followed. From all this it will be seen that spiritual healing is not only very different from any other kind of healing: but that it is also far more inclusive and more wonderful.

HOW CURES ARE EFFECTED

The principle upon which spiritual healing is said to be based is simply this: A certain vital and magnetic energy is contained in every living body. In health, this is large in quantity, and in disease this stock becomes depleted. Ordinarily, the only way to recover this lost vitality and energy is to rest, sleep and take such care of the body and mind that this vital energy again fills and recharges it to the same extent as before. But this is a slow and uncertain process. It is, however, the only sure way we know. Stimulants, etc., which apparently add strength to the body, do not do so in reality; they abstract it faster. When we expend it faster, we are under the delusion that we are "stronger"; but ultimately we are weaker.

In the case of spiritual healing, on the other hand, it is very different. Vital energy is imparted to the system from without; it fills the nerve-centres and literally adds new life to the whole body. These nerve-centres being aroused, the various functions of the body are stimulated in turn, and in this manner the patient is cured.

THE COSMIC, VITAL ENERGY

This vital energy which is imparted by means of spiritual healing is a great Cosmic power, which pervades the whole Universe. It is everywhere; it is back of every phenomenon: "In it we live and move and have our being." It is illimitable in extent and in power; we simply have to draw upon it to the extent we can; and the more we can "draw," the more rapidly do we become well; the speedier the cure. There is no reason to suppose that, if we could "tap" this great Reservoir in the right way, we should not become well instantly—and indeed there are many cases of this character, where, apparently, this has been done—instances of so-called "miraculous cures" being of this nature. We must learn to tap the source of spiritual energy, and when we have reached this inexhaustible fountain, then health and strength are ours!

"HEALING MIRACLES"

The facts of spiritual healing are as old as history. The "laying on of hands" was one of the most ancient modes of treatment, and was employed by the Egyptians. Christ employed it frequently. When a woman touched the hem of his garment, and he "perceived that the virtue had gone out of him," he doubtless felt a loss of the precious vital magnetism, by means of which he effected his marvellous cures. The healing miracles in the New Testament are full of cases of this character; and in our own day we often read of wondrous cures, effected by those who have somehow learned to come in touch with a Higher Power,—some source of energy not available to all of us,—and to draw upon it for the purposes of their "healing miracles."

To a certain extent, doubtless, we draw upon this fund during our sleep; but it may be drawn upon in far larger quantities by those who have the secret of how to do so. Some spiritual healers can do this; but discarnate spirits can apparently direct and manipulate this vital energy far more effectively and to better purpose, for the reason that, living as they do in the world of spiritual energy,—they understand more of its laws, and can better control and govern them. Hence they can effect a cure, very often, when every other means has failed.

POSSIBLE EXPLANATION OF SUCH CURES

While it is true that most cures depend upon this vitalizing magnetic current, it is possible that in certain cases, actual physical transformations are effected. There are many cases on record in which actual tissue has been replaced, apparently instantaneously, by some extraordinary means. In these cases, it is possible that the spiritual energy has actually built up a part of the body, out of matter and the vital forces which were employed,—"materialized" it, in fact—and left this part of the body whole and sound, as before. To those who believe in the reality of materialization—that human bodies of flesh and blood can be built up out of

invisible elements—there is nothing incredible in this
suggestion. But it is only advanced as a tentative and
possible explanation of certain facts which have, to date,
received no explanation whatever.

HOW TO DRAW UPON THE COSMIC ENERGY

The great question is: How are we to draw upon
this great store of energy? If we are alone, how effect
a cure within ourselves? And if you are a medium,
how cultivate and develop the power of drawing upon
this Cosmic Energy, to such an extent that cures may
thereby be effected through or by the means of your instrumentality?

Let us take the former question first. We will sup-
pose you are alone, with no one near to help you.
You desire to be helped and cured by spiritual means.
What are you to do?

In the first place, you must learn how to *relax*. If
your muscles are tense and rigid, you will never receive
any influx of spiritual energy. You shut it out, since
the receptive attitude is the only one in which this energy
can be obtained. So you must insure complete muscular relaxation. It may be obtained as follows: Lie on
a hard couch, with no support for the head. Relax all
over as completely as possible. Then *think* of your
neck. You will probably find it tense and stiff, when
your attention is turned upon it, and that you are holding your head on your shoulders! Relax it—allowing
the head to sink into the couch and support it. When
you have done this thoroughly, think of your right arm,
and relax that; then the left arm, then the right leg,
then the left leg, and finally the whole trunk. After
you have encircled the body in this way two or three

times, you will be well relaxed; and you must then begin your breathing exercises.

Breathe slowly and regularly, inhaling from the diaphragm, not the chest. Breathe through the nose, as before explained. Keep up these breathing exercises for five minutes, expanding the lungs, and seeing to it that you have plenty of fresh air. This will be quite enough for the first day or two, and it is inadvisable to try any more. You will arise refreshed and invigorated, as the result of your exercise.

PROGRESSIVE EXERCISES

On the third day, you may begin your mental practices when breathing. With every breath you take in, think to yourself, "I am power; I am strength; I am health; I am well!" etc. Keep this up for three or four minutes, concentrating upon it, and really believing it. Then rest quietly for a minute or two; then quietly and hopefully call upon this Spiritual Energy to cure you. Remember the more completely you can give yourself up to the influences which come to help and cure you, the more completely and rapidly will you be cured. Send out a mental call for help and assistance, and it will surely come to you!

THE FUNCTION OF THE VITAL BODY

Spiritual healing depends, very largely, upon the fact that the physical body can be acted upon, and influenced, from higher spheres and planes of activity, through or by means of the vital or etheric body, which inhabits the physical body. This inner body acts as a sort of medium or vehicle, through which the cosmic energy flows; and the problem is to connect-up this inner body both with the physical body on the one hand, and with the great reservoir of spiritual energy on the other.

It must be admitted that we do not know exactly how this is done, in the present age of the world's spiritual evolution. If we did, we should be enabled to perform almost "miraculous cures," instantly, resembling those of Christ, who doubtless possessed a wonderful knowledge of these laws. If the law is to be discovered at all it is doubtless along these lines. Experiment therefore; and when you hit upon certain positive results, you may be sure that you have discovered a portion of the Great Truth. Do not assume, however, on that account, that you have the Whole Truth for you will make a great mistake if you do.

HOW TO BECOME A SPIRITUAL HEALER

Now let us suppose that you are a medium, and that you are treating some one else. You desire to gain this power and to obtain assistance from the spirit world. This is how you should proceed.

You must first of all see to it that you are in good physical health. If you are not, your vital magnetism is apt to be tainted and injure the patient. Further, as you often draw the patient's ills from his body into yours, you must be in good health to do this. Next, your mind must be receptive, sympathetic and in an attitude of kindly helpfulness. If you feel selfish this at once sets up a barrier or wall, which you will be unable to break through. Finally, your psychic sensitiveness and mediumship must be developed, to a certain extent, to enable you to practice this "phase" with any hope of success. The methods which you must follow to increase your mediumistic power have been explained in some of the previous chapters, and will be more fully explained in those which follow.

Now, assuming that you have your patient before you. Place your hands on his forehead, and make gentle strokings. Then place one hand on his forehead and one on his solar plexus. Take a number of deep breaths, asking your patient to close his eyes and breathe with you, in perfect rhythm. In this way you get into unison and sympathy. Then make yourself negative, and ask the Spirtual Power to come and help and assist you in your process of cure. Make yourself a *channel* for it. You will feel tingling sensations in your arms, and the patient will feel them in his body. This is the beginning of the process. Try to find just the right mental and spiritual attitude, and power will certainly come. From day to day, your ability to draw upon the great Cosmic Energy will increase. You will get greater and greater power, and as this develops, you will be able to handle and control it more and more. Your power as a spiritual healer will in this manner increase from day to day.

CHAPTER XIX

THE CULTIVATION OF SENSITIVENESS

"SENSITIVENESS" means the ability to sense or perceive in some subtle manner, auras, impressions and influences, either issuing from another living person, or from some thing, or emanating from "spirits." In so far as a sensitive or medium can sense or feel these influences, he is a psychic; and the cultivation of this power is, in a sense, the essence of all true progress in mediumship. This is, therefore, one of the most important Lessons which can be learned, for as you progress in psychic sensitiveness, you also progress in mediumship,—other things being equal.

The first chapter of this book was devoted to Development; but that gave only the "outward form," as it were, of the process; and did not enter in any way fully into its *essence*. We cannot pretend to do so even in this chapter,—since the subject is too vast and too delicate. But I may take the student some distance further along the road,—for, after he has mastered the preceding chapters, he will be more enabled to undertake these exercises than he was at the beginning of his development.

HOW TO DISTINGUISH TRUE FROM FALSE

One of the greatest difficulties, doubtless, in the cultivation of sensitiveness, is how to distinguish true from false—hallucination from reality. At first, this will doubtless be next to impossible, and many false steps will have to be taken before you find,—from actual bitter experience,—what is true and what is not. But as the inner sense becomes developed, you will find that it not only gives you the knowledge in question, but that it also enables you to distinguish one from the other— true from false, and illuminates the whole subject so that mistake is almost impossible. This certainty, which you will then have, cannot be communicated to another; it is often impossible to prove to one who does not experience this inner vision of reality that what you receive *is* true, none the less! As Mr. Charles Brent says, in his *Sixth Sense:* "The serious crux is how, in the realm of the spiritual and the physically intangible, to distinguish between the real and the seeming, the true and the false. This is the function of the Mystic Sense to do, aided by the full complement of inner faculties. In a measure the Mystic Sense, like the bodily senses, acts automatically, but like them it needs *special training* in order to separate phantasm from reality, to determine values, and to grade and classify ideals until they reveal themselves to be ordered unity, not less but more mysterious because more intelligible . . . to the whole man." It is because of all this that long training in psychic development is necessary; and sudden jumps or leaps into full possession of this knowledge is impossible.

THE FIRST STEP

The first thing to do, in cultivating your inner sensitiveness, is to stimulate your physical senses to the point of their highest activity. Endeavour to perceive and feel vibrations unfelt by others,—for much depends upon vibration. Train your senses. Then train yourself in

seeing auras and in psychometry, as before explained,—
in this way getting further along the road. Try to see
and to feel the emanation coming from people you meet;
look at them steadily, and see whether you cannot discover a sort of hot air or vaporous emanation issuing
from their bodies, and radiating out into space. As soon
as you have succeeded in this, begin to analyse your
feelings and emotions, and interpret them. Do this,
(1) when you touch the person in question; (2) when
you receive a letter from him, which you should hold in
your hand— or between both hands; (3) when you hear
him speak; and (4) when you merely see him. When
trying these experiments, assume a "listening" attitude,
as before explained and breathe slowly and deeply.
(This breathing must not be too conscious, so as to take
your attention, however.) Relax yourself as much as
possible during this period. Try in the dark or semi-
dark, at first; in the light later on.

PSYCHIC ATMOSPHERE

When you are walking along the street, cultivate the
practice of sensing persons, and seeing their aura. You
will soon be able to feel a sort of air or atmosphere about
each individual—just as there is a definite air or atmosphere about a house or a town. Thus, a manufac-
turing town has quite a different "atmosphere" from
one which is not. You will soon be able to get this,
in a general manner.

After you have progressed thus far, you should endeavour to feel any cuts, bruises, pains, etc., which may
be upon a person's body. You should do this, at first,
by passing your open hands gently over the surface of
the body, and, as soon as you come to the spot which is

sensitive and sore, you will feel a slight pain in your own body in the corresponding place. Before you are able to do this with much success, however, you should develop certain phases of psychometry,—as for instance the following.

Make a number of small paper packages, all exactly alike in appearance. In these place salt, pepper, mustard, cloves, nutmeg, sugar, cayenne, etc. Mix these all up so that you cannot tell which is which. Now practise feeling or handling these until you can tell the contents of any given package by merely feeling the paper in which it is wrapped. As soon as you have done this, you are ready for more advanced practices.

Having progressed thus far, you are in a position to try your first experiments in psychical diagnosis. Pass your hands over the body of your patient, (who should be divested of as many clothes as possible), and if your sensitiveness has begun to develop, you will feel a pain or some sensation in your hand or arm, or in some corresponding part of your own body, as you reach the diseased spot in your patient's body. Cultivate this until you can succeed with more or less certainty and precision. The more you practise this, the more perfect you will become, and the more rapid your advancement will be.

When you have reached this stage, you must go one step further. Having located the seat of the trouble, and its general nature, you must seek to know how to cure it. Hold the mind in a receptive attitude, when doing this, and you will soon begin to receive the distinct impression that you must do something for the patient— but you will not know as yet *what* it is. After a little time, you will get the distinct impression what to do—

to make certain passes or manipulations, to prescribe a certain drug, to apply certain water applications, etc. As soon as you have reached this stage, you are on the high-road to becoming a successful "spiritual healer," and your power will develop with every sitting. It would be well for you, at this stage, to sit by yourself especially for development in this direction; and added power will doubtless be given you with which to work your cures.

PROGRESSIVE EXERCISES IN SELF-DEVELOPMENT

It may be, however, that you do not care to develop your sensitiveness in this direction. You wish only to develop it for your own progress, and not for the purpose of becoming a healer at all. In that case you must follow a slightly different method of development—though all the exercises we have described will be found helpful and advantageous.

If you desire to cultivate your own sensitiveness, the following exercises will be found very useful in this connection:

1. Try to analyse your own emotions when in the presence of (a) a large company of people, and (b) a small gathering. You will probably find that your impressions are very different, and that a large crowd will give you the impression of being more scattered in mind than a small one. In other words, you will begin to sense or feel the "mind of the crowd." It is well-known that such a thing exists,—for crowds will often do things and perform actions which no individual in it would perform alone. If you can sense this mind of the crowd, your sensitiveness is progressing favourably and rapidly.

2. Stand before a mirror. See whether or not you are enabled to perceive any influence coming from your reflected image in the mirror. Many sensitive persons can do this, and the more sensitive you are, the more will you feel this. You will sense a magnetic fluid, coming from the reflected form in the mirror.

3. If you are in the habit of sleeping with any one regularly, endeavour to analyse the impressions you receive from the aura emanating from the body of the person with whom you may be sleeping. See whether this is positive or negative. Positive aura is slightly warm, negative aura is somewhat cold.

MORE TESTS FOR SENSITIVENESS

4. Hold your right hand above a mirror; then the left hand. Try to feel whether one hand feels cooler than the other; or whether both are of equal temperature. The hand which is warmest is on the more positive side of the body.

5. Close your eyes, and have some one make magnetic passes over your head and shoulders. Try to tell whether those passes are being made in an upward or downward direction. Downward passes are positive or sleeping passes; upward passes are negative or waking passes.

6. Procure several metals,—such as copper, iron, tin, zinc, etc. Place your hands over each in turn, and ascertain the different impressions you get from each one. Then wrap them in separate pieces of paper (making all alike in appearance) and see whether you can always tell the correct metal from feeling the paper. Then place your open hand over the paper, without touching it. Next remove your hand gradually further

and further away, until you are some distance from the metal. After a time, you should be enabled to do all this from a considerable distance. It is only an extension of this power which enables "dowsers" or metal and water-finders to locate the whereabouts of metal and water under the ground, by walking over the spot, above ground, with hands outstretched, or with a divining or dowsing rod held in their hands.

7. Always have flowers in your sleeping room. They are a good influence. Analyse the difference between your impressions when the flowers are removed; and when they are in the room.

COLOURS AND EMOTIONS

8. Procure some water-colours and paint solid strips of colour on a piece of white paper. Make these about half an inch broad and three inches long. On one piece paint a bright red strip, on another a vivid blue, on another emerald green, on another black, etc. Blindfold your eyes, shuffle the papers, and then place your hand on the topmost one, and see whether you can tell from your impressions what colour you are touching. Red will give you a sensation of warmth, light blue of cold, etc.,—as explained in the chapter devoted to "Colour and Its Interpretation."

9. Try to cultivate what is known as Sensitiveness to "psychic contagion." You must remember that thoughts and emotions are just as contagious as diseases; and that you can "catch" them in just the same way! When in the company of other persons, therefore, endeavour to catch or feel their emotions and feelings. You will probably get, at first, the thoughts, etc., they are expressing; then those which they are just about to

express—so that you "take the words out of their mouths." Then you will begin to sense the feelings and emotions of the speaker before they are put into words; finally you will be enabled to appreciate his whole feeling and thinking Self,—by a species of intuition or impression. Endeavour to *draw* this *out* of your subject, and do not let it come to you without any effort on your part. Be *active*, that is, instead of merely passive. In this way lies safety and success.

THE EXPRESSION OF IMPRESSIONS RECEIVED

10. Finally, you must teach yourself to express what you feel. Often this is most difficult. You may feel a thing, and feel inclined to say it, but something seems to hold you back until it is too late. Overcome this restrictive feeling. It is important you should do so, for this is one of the most important things to learn in the cultivation of mediumship. When you have learned to express your impressions, you have progressed far along the Road.

These practices in the cultivation of sensitiveness are the most valuable you can have as a preparation for the cultivation of true mediumship. At the same time, they are *safe* exercises to follow. Practise them, therefore, before you attempt any definitely mediumistic exercises; and you will be rewarded by a safe and sane increase of your inner, spiritual faculties.

CHAPTER XX

TRANCE

TRANCE is a condition into which certain mediums enter in order to receive messages and give them in the form of speaking or writing. No one knows, at the present time, *what* the medium-trance is, or for that matter, any other kind of trance! Dr. George Moore, in his "Use of the Body in Relation to the Mind" says: "Trance is a state of body sometimes produced in man—a condition utterly inexplicable by any principle taught in the schools." Prof. William James stated his belief that the medium-trance was different from any other trance of which we have any knowledge, and this seems to be borne out by the fact that spirit-messages are given in this condition, as well as telepathic, clairvoyant and premonitory messages of all kinds.

WHAT IS "TRANCE"?

Both Trance and Catalepsy occur spontaneously; both may also be induced artificially by hypnotism. Both are mistaken for death, and in many respects they are very similar. In Catalepsy the body is rigid, whereas in trance this is very rarely the case—this forming the chief mark of distinction (external indication) between the two states. What the *internal* differences are we do not know. Various attempts, however, have been made to define them. Dr. Franz Hartmann, e.g., thus distinguishes them: "There seems hardly any limit to the time during which a person may remain in a

172

trance; but catalepsy is due to some obstruction in the organic mechanism of the body on account of its exhausted nervous power. In the last case the activity of life begins again as soon as the impediment is removed or the nervous energy has recuperated its strength." ("Death: its Causes and Phenomena," by Hereward Carrington and John R. Meader.)

When a hypnotist places his subject under hypnotic control, the subject remains *en rapport* with the operator. The influence comes from a living person. In the medium-trance it seems probable that the operator is not a living but a deceased person, and that it is a kind of telepathic influence from spirits which induces this state. In fact, it is brought about by influence from the "other side."

LIGHT AND DEEP TRANCE

There are all grades and degrees of trance, from the very light stage, in which there is but little difference from the ordinary waking consciousness, to that degree of deep trance where the medium is totally unconscious of everything that passes around him. Very deep trance of this character is rare, but many of the most famous mediums have got their best messages while in that condition. The famous Mrs. Piper of Boston had almost to *die*, to all outward appearances, before she could enter this deep trance, and at the end of two hours or so, during which the trance lasted, the only signs of life were slow respiration and heart beat. The only signs of consciousness were manifest in the right hand and arm which did the automatic-writing.

Many test-mediums and sensitives, on the other hand, pass into a stage of trance so light, that no one but an expert could detect any trance at all. Yet, in many such cases, no memory of the condition remains after the trance is finished. These light trances differ but slightly from cases of day-dreams, absent-mindedness, etc., when we say to a person, half in joke: "You are in a trance!" By shades and degrees this becomes deeper, as the state becomes more profound and lower and lower layers or strata of the subconscious mind are reached. Mrs. Piper had three distinct "layers" of this character. The first differed slightly from the waking state. In this condition she talked. The second condition was far deeper trance and in this stage spirits were seen instead of human beings. In the third or deepest stage speech was usually absent and Automatic-Writing occurred.

SPIRIT CONTROL DURING TRANCE

In trance, we may assume that there is a gradual and fluctuating control of the medium's mind and body by the communicating spirit, and that, as one vacates or is driven out by the invading intelligence, the latter is able to control, more and more effectually, the medium. Just as "two solid bodies cannot be in the same place at the same time," so two spirit intelligences cannot occupy and control the same body at the same moment.

When once the fact of spirit control is granted, the nature and character of this control remains to be solved. *How* does the spirit manipulate the brain and nervous mechanism of the medium, to bring about the desired results? What parts of the brain are used, and how? These and many similar questions remain to be answered; and it may take many years of scientific research before we are enabled to answer queries such as these with any degree of confidence.

THE DIFFERENCE BETWEEN SOMNAMBULISM AND TRANCE

The difference between Somnambulism and the Medium-Trance seems to be that, in the former, we remain *en rapport* with ourselves and in the latter we are in touch with the Spirit-World. Many mediums who give inspirational messages or lectures from the platform are in a condition of light trance, and children have been known to pass into this condition and give a large amount of valuable information, unknown to their seniors, and which certainly could not have been known to themselves.

Properly managed, the trance condition is not harmful, though it may become so in the hands of blundering persons. "The Spiritualist's Manual" gives four chief reasons why the trance state should not be harmful to those who enter it. These are:

VARIOUS METHODS OF CONVEYING INFORMATION DURING TRANCE

(1) The intelligences acting upon them (the mediums) are almost invariably of a superior character and therefore must mould the organism by constant use for the expression of higher forms of thought.

(2) The relation of the medium to the manifesting intelligence is that of pupil to teacher, sometimes that of a child to a wise and loving parent, and sometimes both these relations combined with a subtle and ennobling spirituality.

(3) There is always a mutual spiritual relation, even though the medium is not humanly conscious of it; and no one can be a medium for the perfect expression of spiritual messages or discourses who does not consent to the procedure and co-operate with the manifesting spirit.

(4) As the master-musician improves the instrument he plays upon, so also a spirit controlling a human organism for the purpose of expressing wholesome thought, imparts a greater power both to the brain and spirit of the medium. It is often difficult for spirits to control a medium sufficiently to manifest in any way through him.

DIFFERENT TYPES OF CONTROL

There are various types of control which are used by spirits in trance-mediumship.

(a) There is the telepathic method, in which the thought is conveyed direct to the mind of the medium, who is sufficiently awake in light trance, to receive this thought and give it out to the sitter in speech or writing.

(b) There is the picture or "pictographic" method, in which the spirits present certain images or pictures to the clairvoyant eye of the medium, and these pictures are seen and interpreted either directly or symbolically.

(c) There is the sense-impression method, in which general sensations or impressions are conveyed to the medium, who takes on the condition of the communicating spirit, describes pains felt in various parts of the body, etc., as the case may be.

(d) There is the direct-control method, in which the spirit apparently removes the medium's own spirit more or less from the body, in deep trance, and manipulates it as he would an instrument, by acting upon the nervous system direct, in much the same way that we act upon our own nervous systems throughout life.

This latter method is very rare and is only found in cases of very deep trance. Doubtless, there are other methods which spirits employ at times and probably combine all the above on occasion. But these are the most distinctive methods, and they are the ones which may be seen more readily than any others in cases of trance-mediumship.

In cases of so-called "ecstasy" the spirit of the medium obtains the information himself, either by clairvoyant vision or by partially separating itself from the body and visiting the spiritual world direct. The visions which are obtained in ecstasy are thus descriptive of the spiritual world, and what is happening there, and for this reason most of the "Revelations," socalled, are ecstatic visions more or less symbolic.

THE BEST WAY TO ENTER TRANCE

Many persons cannot enter trance spontaneously but have to be mesmerized by another person before this condition is brought about. Even Andrew Jackson Davis was mesmerized for years before he could develop spontaneous trance,—so that he could enter it at will.

This may be one of the best ways for the beginner to begin his trance-mediumship, but you should take care that the person who mesmerizes you is of a suitable temperament and in every way fitted and qualified to do so. If he is not, you are liable to attract to yourself spirits of a lower order, and then you will bring to yourself lying and malevolent spirits and you may induce a case of so-called "obsession." However if the operator is harmonious and qualified for his task, he will not only prevent this, but would see you through more safely than if you entered this condition spontaneously and by yourself.

Trance is very closely akin to some cases of suspended animation, to certain Yoga trance conditions and even

death itself, which has been called its "twin brother."
As we have seen, however, it differs from all these very
widely. Spontaneous trance is doubtless the most commonly experienced and is the one which you should
endeavour to cultivate in yourself,—other things being
equal.

During these conditions—as you develop—many odd
and striking phenomena will doubtless become manifest
to you. If your hand writes automatically, you will
probably note that it becomes more or less sensitive or
anæsthetic, as explained in the chapter on Automatic-
Writing. If speech is induced, as the result of trance,
this may be striking and coherent, or quite possibly mere
nonsense and "gibberish." If the latter develops you
may be sure that something is wrong, and you should
strive to ascertain what this condition is, and correct
it if possible. Here. as elsewhere, you must be careful
and exercise your own judgment and discretion on the
messages received, and not to accept all these as absolute
truth, for if you do, you are likely to be greatly deceived
—especially at the beginning of your mediumship, where
everything is faulty and difficult. The clearer the communications, generally speaking, the surer the messages,
but those coming through what might be called "Am-
ateur-Mediums" are to be trusted only when they have
been verified!

HOW TO EXPERIMENT WITH THE TRANCE

You may experiment with your own trance-condition
profitably in the following manner:—Sit with pencil in
hand for Automatic-Writing, and induce one or more
friends of yours to do the same thing at the same time.
See whether there is any connection between your writ-

ings, when they are compared the next day. In many
instances, where this has been tried, striking coincidental
messages have been received, partly through one medium
and partly through another. They thus tend to confirm
each other and show that the same spirit intelligence is
active and manifesting through both mediums at the
same time, almost,—or one directly after the other.

HOW TO ENTER TRANCE BY YOURSELF

In order to induce trance spontaneously in yourself
you should proceed, more or less as follows:—Begin by
gazing for some time at a bright object, such as a reflected
light, coming from a mirror, crystal-ball, etc. This will
tend to tire the eyes and nerves slightly, and bring about
a dazed condition which is usually the beginning of
trance. While looking at the bright object, breathe
deeply and regularly through the nose and from the
diaphragm, as explained before in Chapter VI. You
must not let this distract your attention, however, as all
the bodily processes should be unconscious. If you have
already practised deep breathing, as before explained,
you should by this time be so far advanced that you can
do so at will without consciously thinking of it.

While looking at the bright object, do not concentrate
or think of anything in particular, beyond keeping yourself conscious and remembering all the time that you
are "yourself," that you are not leaving your body and
that you are not going to become totally unconscious.

During this process the room should be as quiet as
possible, though some monotonous sound such as the
ticking of a large clock, might assist matters. Do not
listen to this consciously, however; abolish all feelings
of fear and all anxiety, as such mental states will

effectually prevent you from entering the trance condition. "Let yourself go" and develop as far as possible.

SYMPTOMS OF EARLY TRANCE-MEDIUMSHIP

You must not imagine that the beginnings of your mediumship will be either profitable or pleasant, because they probably will not be. Nearly all successful mediums will tell you that they have passed through a period, at the beginning of their mediumship, when they thought themselves in danger, and believed that their minds were being impaired for the time being. This, however, passes off as you progress,—provided that you progress along the right road at the beginning of your mediumship, and this you should endeavour in every way to do. If you can consult an experienced medium or, still better, if you can sit with him during your development, or induce him to be present during your psychic unfoldment, things will be far easier for you and far safer than they would be otherwise.

The oncoming of trance is often signified by certain physical and psychical manifestations, which must not alarm you when they appear,—as they sometimes but not always do. Hiccoughs, sudden and spasmodic pains and cramps, a feeling of "all-goneness," nausea, flashes of light, or the sensation of faintness and that everything is turning black before you—these are a few of the symptoms which you are liable to experience during your early development; and, though they may not be pleasant, you had better be warned of them in advance, and not be alarmed when they appear.

Sometimes, however, none of these signs are manifest —only a delightful sensation of falling asleep upon a bed of roses. In these cases, the psychic has developed

himself properly and systematically, and his guides or
controls are also wise and helpful. These are the for-
tunate—but unhappily rare cases; but it is hoped that,
by following the advice given in this book, many more
will be enabled to develop in this wholesome manner.

THE THREE RULES TO FOLLOW

There are three chief and most essential factors to be
considered:—

(1) Your mental and physical health must be quite
up to the standard. If you are depleted, exhausted,
or "run down" physically, if you are suffering from any
disease, or if, on the other hand, you are full of fear,
apprehension and doubts, or if anger and similar
thoughts rage in your soul, you may anticipate a difficult
time in your development and unpleasant experiences
throughout that slow process.

(2) You should be careful to keep your self-consciousness active and alert when entering trance. Do not
give yourself up completely or allow the mind to become
a blank *at first*. Give yourself up in every other way
but this. You must always keep in the background of
your mind the thought:—"I am I . . . I am so-and-so
(your name) . . . I will remain in my body. . . . I am
strength and power . . . I will not be influenced *against
my will* by forces other than good. . . . I can always
return to myself when I want to." These and similar
suggestions you must give to yourself, and hold them in
your mind as a central point of force while entering
trance, even when allowing yourself to become passive
in every other way. If you do this, you will avoid a
great deal of difficulty and danger.

(3) If you can in any way assure yourself that you

have a band of spirits or controls "on the other side" who are ready and willing to help you, this would mean much. A good medium or clairvoyant could probably tell you whether this is the case, and the nature of the intelligences who are trying to influence and act upon you. If these are described as evil, you had best postpone your development until this condition changes. If, on the contrary, they are described as good and helpful, you may proceed, subject to the above precautions and advice.

IMPORTANT CONDITIONS TO BE FULFILLED

It is important to have a plentiful supply of fresh air in the room when entering trance, and after you are in that condition. Also the light should be so regulated as to affect you most agreeably. This may be semi-darkness, though many trance-mediums develop in full light. Soft music may be found beneficial in some cases, though not in all.

You should have everything ready to hand—such as pencils, paper, etc.—before you enter the trance-condition. During the trance-state you will probably be more or less sensitive to objects placed in your hands,—that is, you will be enabled not only to psychometrize them, but, in connection with the objects given you, you will get spirit-messages and information concerning the individuals to whom they formerly belonged. All objects of this character carry the aura or influence of the person with whom they have come into contact, and, for this reason, those objects which have been next the skin, are the best for this purpose: gloves, headbands, etc., are especially valuable. These should be wrapped, as before explained, in oil-silk, and they should be handled as little as possible after the death of the person to whom they belonged.

DEVELOPING EXERCISES

A very good practice in developing trance-mediumship in yourself is to cultivate the habit of analysing your own "falling asleep" process. Try to catch yourself as you fall asleep and hold on to yourself when in this semi-sleeping condition as long as you can, before finally dropping off to slumber. This you will find very difficult at first, but it can be mastered more or less in time.

If you can succeed in catching yourself in this manner, when nearly asleep, and retaining a certain degree of conscious control, you may rest assured that you will not only be a good trance-medium, but that you will be able to protect yourself while in the trance-state, and that harm can hardly come to you when in this condition.

This is a very excellent practice and has given many psychics that power over themselves which they formerly lacked.

Spiritual repose is essential for the trance-medium who would develop simply, harmoniously into practical and wise mediumship. In this manner, you are said to come in tune or harmony with the great Cosmic Currents of truth and wisdom which flow hither and thither in our world, and to and fro from the Spiritual Source of love, wisdom and intelligence. Once get into harmony with this stream, and your progress, not only as a medium, but as a Supreme Psychic, is assured.

CHAPTER XXI

INSPIRATIONAL SPEAKING AND TEST-MESSAGES

INSPIRATIONAL speaking depends partly upon the activity of your own subconscious mind and partly upon the amount of help you receive from the spirit world. In speaking before public audiences for the first few times, you had best think over what you are to say and prepare your talk a little in advance, then depend upon the help and inspiration you receive for the elaboration of these notes, which you have made. As you progress you will find that less and less of this preparation is necessary, and after a time you will be enabled to dispense with it altogether, and know nothing more of the subject of your discourse than the mere title.

ON THE ROSTRUM

When on the rostrum, you can close your eyes and the discourse, more or less eloquent, will flow from your lips. When you are still more advanced, you will be enabled to allow persons in the audience to choose subjects, and you will then be able to talk upon these at great length and often with a profundity of knowledge and beauty of style which surprises not only yourself but your auditors. Many of the best and most profound object lessons and instructions have been received in this manner, and much of the philosophy of Spiritualism has been propounded and explained in this way.

Test-Messages are of a somewhat different order and are given in a different manner. They relate to persons in the audience or to objects brought by those persons, and the information concerns not an abstract subject or theme, but individuals connected with the person to whom the message is given. It concerns spiritual personalities more directly than spiritual truth, and though both may have their origin in friendly helpers, they are nevertheless given for different purposes and in a different manner.

HOW TO BEGIN

A very good way to begin training both for Inspirational-Speaking and Test-Messages is the following, which Dr. R. R. Schleusner—an able trance and test medium—personally followed in his development, and which he found gave most satisfactory results.

In writing of this, Dr. Schleusner says:

"First of all I ask some one in the audience to speak to me a set or formal sentence, such as: 'Doctor, help me,' or 'Doctor, reach me.' From this I receive certain impressions which I analyse somewhat as follows:

"1. I can tell from the sound of the voice whether it is harsh and grating or whether it is soft, gentle and harmonious, and from this the character of the speaker may be more or less diagnosed. Also the voice will tell me whether the person is nervous and irritable, or self contained and controlled: whether the person is angry, or is sceptical and merely asking for a test. Further the voice will tell me whether the person is weak or strong, positive or negative, sensitive or the reverse. These are the physical properties of the voice, so to speak, and from them I gather certain information more or less instantaneously and subconsciously as to the sitter and his attitude.

COLOURS AND AURAS

"2. In addition to this I receive, in connection with his voice, certain psychic impressions. These take the form of auras or colours drawn up in cloudy pillar-like forms. These colours I interpret symbolically (according to the interpretations outlined in the chapter devoted to Colour and Its Interpretation). Thus if I see before me a dirty slate-grey I say that the conditions before me are at present very unfavourable and depressing, and if, just beyond this, I see a yellowish golden rim, I state that the immediate future prospects will be much brighter and better and that the person in question may cheer up, as better conditions are coming, etc.

"Besides this colour, which is drawn up before me in this form, as the result of the physical vibration of the voice, and which seems to be caused by it, I always see another series of colours and auras in another place, some distance from the first in space, which I compare with the former set and see whether or not they agree. After seeing the first set of colours I close my eyes for a moment, then open them and look at the second set. If they are found to agree with one another, I know that my first impressions are correct, and I then state openly that such and such conditions are so. My own experience has been that if these two sets of auras agree with one another (those which are produced by the voice and those which are apparently shown me by my spirit-guides) the diagnosis or psychic impression is correct, and I am very seldom wrong in my statement of the fact.

THEIR INTERPRETATION

"Let us go back for a moment to the impression received by the heavy leaden grey colour. Suppose this

is the colour seen; this indicates depression. Having
arrived thus far, the question is: how to get out of this
condition. That is suggested partly by common sense.
Having proposed this question to myself I close my eyes
and look at a different place in space. In this third
place I then see presented to my psychic sight a symbol.
This symbol tells me how to escape from the present
difficulty. After I have explained the symbol, and interpreted it to the best of my ability, I then look back
at the colours and see whether or not they have changed.
If they have become brighter, then I know that this
is the correct path to follow, and that good will result
from the course of action advised. If there is no change,
I state that things will continue for some time to come
in this depressed condition, and that the best that can
be done for the time being, is to hope and work on
patiently.

HOW TO INCREASE YOUR POWER

"These auras, colours and symbols may be impeded
or shut off by certain psychic conditions on the part
of the sensitive. For example: You may hold on to
them too tightly, as it were, and this tenacious grasp
for too long a time will have the effect of shutting them
off altogether. You must learn to let go as soon as the
symbol has been perceived, or the colour seen.

"As these colours are presented to you, you may, however, see a change going on, and in that case you should
watch it intently and see what the change may be. Thus,
grey may change to white, as a sign of spirituality, and
you can state that the person is becoming more spiritual
and changing his point of view in life, clinging to his
ideals, etc., and that if he continues to do so success
will reward him.

"If you see tinges of golden yellow, you may be sure that the individual in question is cultivating his intellect and that he is independent in thought and a more or less clever, intellectual person.

GETTING HELP FROM SPIRIT GUIDES

"In addition to these colours and symbols, other phenomena may occur. Thus, in my own case, I always see my guides who stand by my side telling me what to say. If I speak aloud just what they tell me, it is usually correct. If I endeavour to elaborate or extend it, it is often wrong. This is an error which should be avoided. Thus: they may say 'three.' This may mean three minutes, three hours, three days, three weeks, three months, three years, etc. At the time I hear the figure I do not know what its meaning is. I therefore say to the person receiving the message: 'My guides say to me "three"' and I then wait for further information.

"As explained in the lesson on 'Symbolism,' the difference between the two consists in distinction between impression and expression. I receive the impression correctly but must be careful not to give it wrong expression. What one should do, therefore, is to wait for further impressions before expressing anything. In such a case as the above I would, after hearing the word 'three' spoken, turn to my guide and ask what the three signified. On receiving an answer, I would state this also, and then go back for further information, etc. If you proceed in this manner you will rarely go wrong.

"The more anxious you are to receive psychic impressions and to give tests, the more fluctuating or changing will your impressions be. The colours of the

auras will keep changing, and often you will see them
constantly varying around any object placed on your
table, whereas if you were in a calmer frame of mind,
these colours would appear stationary. Endeavour,
therefore, in every way possible to control your apprehension and anxiety in giving public tests of this nature.

IMPORTANT RULES TO FOLLOW

"One very good method in giving tests to the public
is to endeavour to force your clairvoyant perception
before asking aid of your spirit friends. Thus, suppose
some member of the audience asks the question 'How
is my mother?' Instead of waiting for a direct impression, in this case I should say in reply to this: 'I
will first try to find your mother and describe her. If
I can do this I will proceed with the test.' This is far
more satisfactory both to the speaker and to yourself.
I now force myself, so to speak, clairvoyantly; and
generally in a few moments the form of a lady arises
before me which I describe to the speaker in as much
detail as possible, giving also her surroundings and the
description of any other persons I see with her at the
time. If this description is incorrect and not recognized,
I then ask whether the description suits any other person
known to the speaker—any person through whom the
mother might be reached. If they reply in the affirmative, I then endeavour to find the mother clairvoyantly
and, if I cannot do this, I ask my guides to give me any
information they can regarding her, without this vision.
If, on the other hand, the test is not recognized, I drop
it and proceed to another case, as I know that I can
get nothing definite for this individual. Instances
of this character are, however, very rare, amounting

probably to not more than 5 per cent. of the tests I give in public.

HOW TO RECEIVE IMPRESSIONS

"In all inspirational and test-messages, you should throw yourself, as completely as possible, upon your spirit friends after you have once asked their assistance, and should be as responsive as you can. Do not wait for them to hammer any impression into your head before you state what it is, but hold yourself rather in the attitude of an empty vessel, and imagine a funnel in the top of your head into which ideas and impressions of all kinds are being poured. As they enter your mental and psychic life you should interpret and express them as best you can."

The foregoing is an exact account of Dr. Schleusner's method of delivering Public Tests, a careful study of which will doubtless prove helpful to the earnest student; and, I believe, will result in a corresponding degree of development in all like cases.

CHAPTER XXII

MORE AND LESS DEVELOPED SPIRITS

INASMUCH as we are said to be "spirits" here and now, —just as much as we ever will be,—we should begin the course of progressive development here in this life, which we intend to follow later. Anything we learn here will doubtless help us in any future development whatever that may be. Wisdom, knowledge, understanding, sympathy and penetration of perception will all help, assist and advance us, no matter *what* world we may inhabit or whatever its nature may be. This being the case, we should endeavour to develop our own inner nature here and now, as suggested especially in the chapter on "Self and Soul Culture," and we should continue this, so far as lies within our power, even after we have crossed the "Great Divide."

WHY STRANGERS OFTEN COMMUNICATE

Just as there are all kinds of characters and natures in this life, so there are said to be individuals of all kinds in the next; and the unfortunate part of it is that many of those who can most easily come back and communicate, are those on the lowest rung of the ladder,— those who are the most earth-bound and belong to the lowest strata of society. They are nearer the earth than more advanced spirits, more in sympathy with it and its vibrations, and their character, being more earthly, is naturally more open to receive and send messages than those who, when alive, had less sympathy with

the earth and felt less bound to it. It is for this reason
that "strangers" often communicate with us more easily
than our nearest and dearest,—than our friends or relatives. The latter may feel for and with us most keenly,
and may long to communicate, but often obstacles hitherto undreamed of may prevent them from doing so.
They then find, for the first time and to their aston-
ishment, that the difficulties of communication are so
great that they are unable to send messages, however
much they may desire to do so, and even if they can
find a medium suitable to receive them.

THE CONDITIONS FOR COMMUNICATION

You, on your side, may be receptive to these conditions, and again you may be unfitted to sense anything
of the kind. The combination of circumstances for the
transmission of spirit-messages is rare, and, in order
for these messages to be clear, as we should desire them
—there must be an effort both on this side and on the
other, made at the *same moment*, also there must be the
"medium." As this combination is naturally lacking,
in the majority of cases, you can see why it is that authentic cases of spirit return are, comparatively speaking,
so rare and why it is that more persons do not "come
back." No doubt you have often heard the objection,
that, if Spiritualism were true, out of all the millions
of people who have died, wishing to communicate, there
must be many thousands who could return directly and
state clearly what they wished to send! What we have
said above will explain the reason why this is not the
case.

The ability to transmit a message from the other side
may be as rare as the ability to receive it on this side.

Good communicators may be as rare as good artists, painters or sculptors. It may be a special faculty which we have to develop, and, just as uneducated and ignorant persons often possess extraordinary gifts and talents in certain directions, for no reason we can see, just so certain individuals may become good mediums or communicators, after they have passed over, simply by reason of their psychic constitution or make-up. The question of the "difficulties of communication" will be found fully discussed in Chapter XXIX.

WHY LOW OR "EVIL" SPIRITS COMMUNICATE

It is because of all this that we often reach or come into contact with persons of a low order, in spirit-communications. Mediums believe that there are tramps and "hoboes" on the other side, just as much as there are in this life. By nature and by instinct they remain the same, and they have to be gradually educated and trained in order to outgrow their natural instincts.

And, just as these tramps and hoboes would be insulting and often disgusting in this life, and would swear, curse and do other things unsuitable for the family circle,—were they introduced into it,—just so they do the same things when they communicate, and get once more into the "earth-atmosphere."

These are the characters who also harm mediums unconsciously by rough handling, so to speak; and by damaging the delicate nervous organism upon which they operate, when sending messages. The best and safest way to guard against personalities of this character, we are told, is to call to your aid spirit-controls, "guides" or advisors who can assist you from the other side, by arguing with such personalities, and by removing them from your aura more or less forcibly, should the occasion demand it. There are many cases on record in which more gentle measures did not bring about the required result, and, according to accounts received, very forcible methods had to be resorted to, in order to eject these strangers, before peace and harmony were finally restored!

"HIGH" SPIRITS AND THEIR HELPFUL MESSAGES

When "spirits" of a higher order come, all this is reversed. You then come into contact with spiritual natures, and help, comfort, sympathy and sound advice are given. When once you are assured of the assistance and co-operation of one or more individuals of this character, your time of tribulation as a medium is more or less over, and thenceforward you may depend upon steady and harmonious progression and advancement in your mediumship.

You must be careful, however, as to how you receive messages claiming to come from exalted personages, as great names will often be given when the individuals in question are not there at all. This may be due at times to accident and misunderstanding; but there is also evidence, unfortunately, that lying spirits will resort to this stratagem to gain your confidence. You should, in this case, rely on your own common sense and judgment, and insist upon proof of identity and direct evidence before you believe that the individual in question is really there.

GUARDIAN SPIRITS

There is one "sect" or division of spirits whose office and general work and interest is particularly helpful to mortals, and that is the so-called "Guardian Angels" or Guides who help, govern and advise friends of theirs still in the body. The sympathy and counsel offered by these Guardian Spirits is, at times, very great; these spirits are said sometimes to prevent accident, suicide and even murder by their kindly help and assistance. It is one of the most beautiful and inspiring thoughts in the Spiritualistic Philosophy to believe that those we love are constantly about, helping and cheering us, along our hard and narrow way; and that they see our trials and tribulations, and share them with us, just as they did on earth. We must feel, too, that they are preparing a place for us and that, when it comes time to solve the Great Mystery, we shall find helpful and loving assistance, instead of a foreign land, into which we shall then enter.

WHO MAKE THE BEST COMMUNICATORS?

Those who possess a simple, open, candid child-like nature, are doubtless those who make the best "communicators,"—other things being equal. It is because of this that Indians, who lived close to nature, so constantly communicate and act as "guides," and doubtless for the same reason Negroes are very psychic, and receive many psychic phenomena.

There is a great deal of evidence, also, as we know, to show us that animals perceive spirits and psychic manifestations, and that they also "sense" phenomena more keenly than human beings. Between animals and ourselves there is doubtless a link which unites us all into one conscious whole,—this being the life of the universe which runs through every sentient thing.

It is not uncommon for "spirits" to return at séances

and seek the prayers or the help of the living. They express themselves as being in trouble and as requiring assistance before they can free themselves and proceed on their way. This is doubtless an important mission to fulfil, and when any wandering and distressed spirit makes itself manifest in this way, it should certainly be assisted in every way possible in the fulfilment of its desire and the discharge of its burden.

Many cases of so-called "haunted houses" doubtless exist, because of the persistent inability of the returning spirit to make any one present see its wants and attend to them. Were a good psychic or medium introduced into such a house, who could get into communication with the returning spirit, and, when communication had been established, help it,—there is no doubt that the "haunting" would cease and that the returning spirit would be greatly helped in its progress and advancement.

HAUNTED HOUSES AND "PACTS"

All this is especially important in those cases of so-called "pacts," where an agreement is made before death to appear afterwards, if that be possible. Many cases of this character are on record, and, whenever such an agreement is made, it is most important that the living person on earth should fulfil *his* part towards the fulfilment of such a plan. By doing so he may assist to an extent he perhaps does not realize in freeing the spirit's mind from earthly ties and conditions.

POSSIBLY UNCONSCIOUS MESSAGES

In considering this question of returning "spirits," one final and important fact must not be lost sight of, and that is: That messages may often be given through

a medium, or directly, of which the spirit himself may
be totally unconscious. He may think or dream or
visualize a certain thought or message, and this may be
reached or "sensed" by the medium, and given forth
as a conscious and intentional message. The reverse of
this, however, is true. The mind of the discarnate spirit
has been read by the medium in trance. His "mental
pocket has been picked," and he has given nothing voluntarily. Further, his thoughts may have been reflected
upon a sort of psychic mirror, or atmosphere, and there
seen and interpreted by the psychic. This, however, is
a difficult question which will be discussed in a later
chapter. For the present it should be borne in mind
that all messages given by mediums need not necessarily
be direct or intentional. They may merely have been
obtained indirectly from the person in question, and
would not be at all the message he would send, were he
aware of the fact that he was transmitting one. It
is because of this fact that many of the messages appear
to us so trivial and inconsequential.

CHAPTER XXIII

OBSESSION AND INSANITY

IN this chapter, I desire to place before the reader in a fair and clear manner facts which are too often neglected by Spiritualists, but which those who develop or become mediums are apt to find out to their cost later in their development, unless they are aware of the facts at the beginning. "Truth is always best," and it is accordingly best for the student to know the dangers and difficulties attendant upon Spiritualism as well as the bright side. We do not wish to alarm or divert from interest any student by this and the following chapters. Precisely the reverse. But inasmuch as "forewarned is forearmed" the student should be familiar with all the possible risks he is running; as there *are* such risks if he does not develop his mediumistic power systematically and along the right lines, as so often pointed out before in this book.

THE REALITY OF SPIRIT OBSESSION

As we saw in the last chapter, it is often easier for the low and less developed spirits to come in touch with us than those more highly developed; and this is especially the case where mediumistic development has not been such as to bring the student in touch with the higher forces and intelligences. Modern science does not accept the doctrine of spirit-obsession as true, claiming that the cases of so-called spirit influence are really only cases of diseased mind and body, requiring for their

cure proper medical attention. Experienced Spiritualists, however, know that while many cases of apparent obsession may be accounted for in this manner, there
are also cases of real influence coming from less-devel-
oped, disembodied spirits, and as great a psychologist
as Prof. Wm. James said shortly before his death: "The
refusal of modern enlightenment to treat possession as a
hypothesis to be spoken of as even possible, in spite of
the massive human tradition based on concrete experience
in its favour, has always seemed to me a curious example
of the power of fashion in things scientific. That the
demon-theory (not necessarily a devil-theory), will have
its innings again is to my mind absolutely certain. One
has to be 'scientific' indeed to be blind and ignorant
enough not to suspect any such possibility." Dr. L.
Nevius, after an exhaustive study of the cases of demon-
possession in China, and after an examination of contrary theories, stated his conclusion that genuine cases
of obsession were to be found, and that they could not
be accounted for by any other theory satisfactorily.

EXAMPLES OF OBSESSION

Take, again, the case recorded by Dr. J. Godfrey
Raupert in his "Dangers of Spiritualism." He gives
the case of a friend of his, M., who, after attempting
automatic-writing and obtaining it successfully, was unable to leave off the practice when he desired to. Even
at night, when retiring to rest, M. had habitually placed
paper and pencil by his bedside in order to be able to
write at once, when summoned to do so, and he had
frequently been awakened from his sleep for this purpose—much to the detriment, of course, of his mental
and physical health. After this, there had come a still

further development of the mystical power of writing; the pencil, too, had been discarded and M. had begun to trace the writing with his finger in the air. He could thus, it appears, write out a message at some length and was fully able to read it afterwards, just as though there was a piece of paper suspended in the air. . . . Things had thus gone on for many months, when M. at last awakened to the fact that a great transformation was passing over his moral and intellectual nature, and that some other mind had permeated his entire being, and he was now conscious that he was ceasing to think his own thoughts; in short, there could not be any doubt that fetters were being woven around him, which he was growing daily more incapable of breaking. The condition of servility and submission which the control at first effected, was now thrown off and the latter showed signs of absolute power. No treatment, either hypnotic or medical, had the slightest influence upon the strange phenomenon, and M. had now given up all hope from this quarter. Some of the authorities, whom he had consulted, did not believe in obsession or possession. Others ascribed it to hysteria and fixed ideas—help there was none. Dr. Raupert goes on:—

ITS DANGERS

"I tried to argue with the personality and proved to him that he was merely a subconscious product on the part of M. When I persisted in denying the presence of a personality other than and different from that of M., a very frenzy seemed to shake the frame of M. and words of the most abusive kind were levelled at me: "What fools you are," it exclaimed, "to tamper with things you do not understand, to facilitate the invasion

of spirits and then to deny that they exist, to play with
hell-fire and then be surprised that it hurts and burns!
I challenge you to propose any kind of experiment to
test my utter and entire independence of the person of
this idiot, with whom I can do absolutely as I please.
See, how I can handle him and ill-treat him. I am now
beating and hurting him and he can do nothing to defend
himself.'' With this there appeared red spots in different parts of M's. face and he groaned as if in physical pain. Upon this I replied that I should accept
the notion of an independent intelligence, if it could
be shown to be a fact and could be clearly demonstrated.
This he promised to do.''

Many similar cases can be found in this author's works,
particularly ''Modern Spiritualism'' and ''The Supreme
Problem,'' and though they are doubtless coloured to
some extent by the author's religious prejudices, they
are nevertheless valuable as records or ''human docu-
ments,'' and should be studied as much.

There are other Spiritualists who have written much
on this subject of spirit-obsession, as for instance, Dr.
J. M. Peebles, whose work ''The Demonism of the Ages
or Spirit Obsessions'' should be read by all interested
in Spiritualism. Many cases are given in this work.

HOW OBSESSING SPIRITS GAIN CONTROL

Madam Anita Silvani, who is an experienced and cautious medium, makes the following wise statements, concerning the possibilities of obsession in those roughly
or over hastily developed: ''As to the evils, experienced by some persons, who have sat in circles for development or for the manifestation of psychic power I would
say that the whole theory of magnetic control rests upon

a condition of mutual receptivity, being established between the members of a circle, but few reflect that the blending of magnetisms with those who form the spirit-side of that circle, is no less a part of the process, and that without the aid of the magnetism of the sitters present nothing belonging to the spirit-side of life would be obtained. Now in forming a circle how are you to insure absolute freedom from the influence of the low or evil, earthbound spirits, who crowd the streets of a large city? The magnetic aura, created by the circle, hangs in a cloud around it, and draws spirits toward it, even as a magnet draws iron and steel and everything bright and rusty—useful tools and dangerous weapons—will be attracted by the powerful magnet."

POISONED MAGNETISM

If you once admit that the aura of a pure and good person can, under certain conditions, be poisoned by absorbing the tainted mixture from a mixed circle of all sorts of mortals and spirits, you also admit that the good persons can carry home with them a sufficient portion of that poisoned magnetism to form the nucleus of a magnetic state, congenial to the low and depraved spirits, and into which any of them can enter a second time without the aid of the circle. For this reason we are opposed to mixed or miscellaneous circles, especially when sitters are not sincere and known to one another. We believe that possession is not always evil, and indeed it is often necessary. But it is the *continued* control of a highly sensitive medium which does the harm by absorbing his finer life essence. An earthbound spirit is like one who belongs to neither earth nor heaven, nor Gehenna. He has lost his hold on the earth life and has not yet attained to the spirit world. He lives in his astral body and having nothing of his own, must borrow from those both above and below him on the ladder of development.

OBSESSION VERSUS MEDIUMSHIP

Mediumship is necessary! Without it there would be no means of knowledge, no instruments through which to study the psychic plane; but mediumship, in exact proportion to the magnetic powers it confers, becomes a greater and ever greater source of danger, the further its development is carried, unless the control of those powers can be handled with a firm hand and understood in *all* its aspects. Knowledge is the best safeguard, and knowledge will be best obtained by those who can study all the conditions of psychic development.

It is said that there are two forms of magnetism, the astral and the physical. The fundamental difference between them is due to the different conditions under which the astral plane and the physical plane function.

HOW "SPIRITS" INFLUENCE US

It must not be thought, however, that all I have said on obsession relates entirely to Spiritualism or to development in circles or in private. We live all the time in a spiritual world as well as in a material one, and hence are open to the possibilities of obsession or influence, both good or bad, and many show in daily life the fact that this influence is strong, for or against.

Invisible intelligences are said to be with us much of the time,—some urging us on to false and wrong deeds, others helping and encouraging us in actions of kindness, sympathy and benevolence. It is our duty to get in

touch with the latter as much as possible, and then we
shall receive inspiration and enlightenment from higher
sources than any at present about us. The difficulty is
to know how to do this without risk, for, as St. Paul
said, we must "try the spirits" and endeavour to prove
to our own satisfaction whether they are good or bad.

There are various types of obsession, but for our
present purposes we shall omit many of the odd and
exceptional phases, such as vampires, which will be
discussed later on in this book, and shall speak only of
the ordinary type of spiritistic obsession.

THE "MAGNETIC LINK"

The body and mind are doubtless connected by a sort
of magnetic link. The mind and the physical body are
connected by means of a fluidic or etheric body (in shape
the double of the physical body). It is owing to the
fact that this body becomes detached from the physical
frame, at times, that many of the phenomena of obsession and insanity occur. The lines of force are broken,
and the etheric body becomes first of all "loosened"
inside the physical body, and then separated more or
less altogether from it, without the wish of the subject,
who may even be altogether unconscious of the process,
and not know what is going on within him. He only
experiences the resultant phenomena, and it is for this
reason that he does not know what method or course of
treatment to pursue in order to get better or become
cured. All ordinary treatment is, for this reason, useless. Medical and physiological treatment, for the
reason that it acts only on the physical body, not on the
mind; and hypnotic or other psychological treatment is
almost equally useless, for the reason that it acts on the

mind without reaching the physical body. Any form
of treatment which really *cures*, must aim to act upon
the etheric *link* or connection between mind and body,
and to act upon it in such a way that it will become re-
adjusted both to the mind and body, and, this once accomplished, the mind will be restored to its condition
of health and sanity. One of the chief things to do,
therefore, is to act upon this magnetic link, and draw
back the etheric body into the physical.

We know that anæsthetics of all kinds tend to drive
out or disconnect the etheric from the physical body,
and it is possible that some day in the future, science
may discover a drug which will have the reverse effect,
of driving or drawing back the etheric into the physical
body. When this has been discovered, it will doubtless
be the means of curing many cases of insanity at present
held incurable. For the present, inasmuch as this drug
has never been discovered, we must resort to magnetic
and mesmeric treatment and other methods of cure, to
be enumerated more fully later on in this chapter.

EARLY SYMPTOMS OF OBSESSION

First of all let us consider some of the typical
symptoms of obsession when they occur:—

One of the primary things which will be noticed,
probably, will be that the patient will be unable to sleep
properly. He suffers from insomnia, coupled with rest-
lessness and irritability. Soon after this he begins to experience a dull ache or pain at the base of the brain,
sometimes at the base of the spine also. These spots will
be tender to the touch. A general debilitated or run
down nervous condition will be present,—perhaps unno-
ticed until attention be drawn to the fact. If the subject

has been practising automatic writing, let us say, he will begin to have a more or less persistent desire to write. This will keep pressing him forward and urging him to "try what he can get" with pencil or planchette. Thoughts seem put into his head, ideas, impressions and impulses which urge him to perform certain actions or do certain things. These will increase in intensity and frequence.

DANGER SIGNALS

From this point onward, great caution should be used, as the danger point or dividing of the ways has now been reached. A careful student of the occult might point out that the symptoms mentioned above, and in the first chapter, are alarmingly similar to those in the early stages of some types of insanity. This is true! I described them earlier in this book, be it understood, not as *desirable* symptoms, but as those which are *likely to appear,* and for which the student should be on the lookout. The facts should be placed before him and when he is in possession of the knowledge concerning them, he will know how best to meet them, if he observes such symptoms.

We see, therefore, the importance of careful and systematic development in the cultivation of mediumship, as I have so often pointed out before in this book.

After the above stage has been reached, it is probable that the student, who is on the wrong road, will hear words as though inside his head, or externally in space, or possibly in his solar plexus,—though this is more rare. Or the phenomena may take the visual turn, in which case the patient will "see things," mostly of an unpleasant nature, such as snakes, devils, or monstrous or grotesque living objects. Thenceforward, unfavourable symptoms will probably develop rapidly until the patient is completely obsessed and under control. The line to be pursued in cases of this character is twofold. First prevention, second cure!

PREVENTION OF OBSESSION

Prevention. Sound physical health is essential for all wholesome spiritual and mediumistic advancement, and if the patient is sick or ill and especially if run down or depleted nervously, he should stop all mediumistic practices until he is again restored to health. Plenty of outdoor exercise of as rugged a nature as possible would do wonders in cases of this character. Fresh air, both day and night, is essential. Tea, coffee and alcohol should be avoided. Plenty of milk should be drunk by the patient, as this both restores and builds up the nervous system in a way that nothing else can. Above all, plenty of sleep must be obtained, and, no matter whether the patient desires ten hours or more at night, this should be allowed, and plenty of rest at all other times. This is very essential at this stage of the proceedings.

The mental and spiritual health must be maintained equally with the physical. Your critical judgment and common-sense must be exercised now as always, both in judging the messages received and in the general conduct of life. Do not believe everything which is told you, as there are many lying and deceiving intelligences as well as useful and good ones. Above all, do not act upon or obey messages which do not strongly appeal to your own good sense and worldly judgment. If you keep up your mediumistic practices, sit only a short time each day (not more than 15 or 20 minutes at the longest) and, if possible, at the same time each day. These two rules, as before pointed out, are very important. Never allow yourself to continue beyond the time-limit you have set yourself, no matter how interesting the communications may be, but say in a firm, loud voice, "We must stop now," or "I will sit again at the same time tomorrow for the continuation of the message." You should then discontinue the writing.

IMPORTANT WARNINGS AND PRECAUTIONS

Never go inside your own head and examine its processes or introspect for too long a time. The wonders of brain and thought may appeal to your imagination, but never let them influence you or cause you to turn your thoughts inward in the attempt to solve them. If you do, it is sure to end disastrously, and there is no more reason why you should be conscious of your thinking apparatus than of your digestive or circulatory apparatus, which is equally mysterious and wonderful. Let them go on by themselves without thinking anything about them, but using them rather as instruments for your life purposes.

Always keep an interest in external things and live, as it were, outside your head, in the outer world all the time. Become interested in matter-of-fact and worldly objects and interests, as this will tend to distract your mind from yourself and restore you to a condition of normal healthy-mindedness.

Cultivate a sense of humour, and never take yourself or your mediumship so seriously that you cannot see the humorous side of a situation when it may arise. Endeavour to harden the inner self, so to say, and focus and concentrate it at a given point which is under your conscious volition and control. Keep the centre of consciousness always intact and be sure that the centre is yourself!

Do not focus either the sight or the interest on external images or impressions when these become persistent. When this is the case, force yourself to banish them by an effort of will, and by deliberately turning the attention into other, more practical, directions.

FURTHER ADVICE

When dropping to sleep, always keep your mind centred on yourself, and never allow yourself any flights of imagination, and never wonder what is going on or endeavour to catch yourself falling asleep, as you might with safety do at other times. Value sleep and look upon it as a kindly friend. Even those who are seriously obsessed are safe when they are asleep, and no matter how terrifying their daily experiences, voices or visions, may be, they very rarely have unpleasant or terrifying dreams. Sleep is the "resting-time of the soul," and if the spirit is *en rapport* with itself, as it should be, it will be protected from all external influences during the hours of sleep.

Cultivate your own force of will and self-possession. If the emotional nature is too intense, this should by all means be calmed down, especially before going to sleep.

A prolonged warm bath will have a very good effect in such cases.

One other preventative practice will be found very useful. It is this: Seated in a dark room, concentrate your will upon the outer limits of your aura—that is the "auric egg" which was described in the chapter on the aura. By proper concentration and practise you will be enabled to harden or toughen the outer shell, as it were, of this auric egg,—rendering it impervious to extraneous psychic or spiritual influences. This you should always practice before dropping off to sleep. Those who have developed this power in a proper and adequate manner, are absolutely impervious to any evil or malign influences from without.

THE CURE OF OBSESSION

Cure. Supposing now that you have not taken these precautions in time and that you have become actually obsessed for the time being, or that you meet one who is unfortunate enough to be in that condition. What is to be done?

The advice which was given under the last heading should be followed here to some extent. The physical health should be built up by all means in your power. Sleep is absolutely essential and as much of it must be secured as possible. It may be necessary even to resort to sleeping draughts or powders in order to secure the necessary sleep. These should be prescribed by your regular physician, however. While drugs are doubtless harmful, bromide and similar medicines can be taken with benefit at such a time, since the evil effects of the drug are more than counterbalanced by the benefits derived by mind and spirit. Alcohol must be discontinued at once and a milk-diet substituted. You must impress upon the patient (for such he is now) that no one can help him beyond a certain limited point; he must help himself. The cure must come from within rather than from without.

HOW TO USE THE MIND

Distract his mind by all possible methods, so as to make it objective instead of introspective. Do not let him go inside his head for a moment to listen to the voices or to see the visions which float before him; but, immediately anything of the kind occurs, distract his attention and interest him in something which is going on about him, and of as dramatic a nature as possible, so as to insure attention. Do not let him go inside his head, for the more he lives within himself, the more difficult will he be to cure. You must teach him to disregard or to deny the voices, or impressions which insistently come to him. If they flatter him and tell him to do certain things, teach him that these voices are evil and lying, and are not to be trusted; for if they were otherwise he would not be in his present condition.

Never imagine for a moment that an obsessed person is illogical or is not open to reason. His reasoning faculties are often keenly alert, and have often to be appealed to, to effect a cure.

These mental devices are very important, though they do not, as before said, go to the root of the matter.

DIAGNOSIS AND TREATMENT

Clairvoyant diagnosis is very valuable, and a trained clairvoyant can sometimes see the obsessing spirits and describe them. When this is the case, the problem of cure becomes more real and more apparent. If the patient believes in the efficacy of prayer, this is doubtless a potent method of cure. The religious nature is one of the most fundamental sides of our character, and is a factor which is capable of exerting an immense pressure when properly brought to bear. Encourage the patient, therefore, to pray, if he is at all of a religious turn of mind.

Magnetic treatment, such as passes etc., is often very valuable, and will assist in restoring the patient to health, by acting upon the etheric body *direct*, as before explained. Combined with suggestion, this is one of the most potent weapons that can be used in our present state of knowledge. By these methods, also, we can in some cases toughen the outer protective auric shell, if the patient is unable to direct his mind sufficiently to do this himself. By proper striving, however, he will soon place himself in the direction of the great healing Cosmic Currents, and when once he has done so will begin to improve immediately and make rapid progress.

SPIRITUAL TREATMENT

The most important remedial measure must now be described. Inasmuch as the obsessing intelligences are "spirits," usually of an evil or lying nature (though they may be only ignorant or bungling) who have wrecked the medium's nervous mechanism, through their ignorance of how to operate it, they are capable of being reached and removed by other spirits,—that is, we should approach them not from the physical or even the mental plane, but from the spirit side of life. One of the best ways to do this is to secure the services of a reliable trance—or clairvoyant medium, who, in the mediumistic condition, is capable of discerning the spirits, and influencing the patient. An experienced medium of this character has with him certain "guides" or "controls" who are helpful and friendly. These guides, if called upon, will assist by arguing with, or, if necessary forcefully removing the obsessing intelligences who are influencing the psychic. The case should be explained to the medium's guides, when the latter is in trance, and their assistance asked. They will then undertake to remove the obsessing spirits, and will often succeed in doing so after a few trials,—though in some instances they are unable to persuade or induce the obsessing spirits to leave, and are not powerful enough to enforce their removal.

This, however, is the most potent and effectual method which we know at present, and should be employed whenever possible. If a medium of this character is unavailable, then a second person should speak to the obsessing intelligences direct, and reason with them as he would with a human being,—pointing out to them the uselessness of the proceeding, the injury they are doing the medium, the harm to themselves, etc.

It is rare indeed that such measures, properly applied and coupled with the mental and physical treatment, described above, will fail to remove the influences which are at work.

THE BRIGHTER SIDE OF THE SUBJECT

I have described fully and freely the "seamy side" of Spiritualism, and its possible dangers. The student must not be discouraged, however, from this black outlook.

The dangers exist, truly, but they also exist in other lines of experimental research, and many lives have been lost in attempts to perfect some system of medicine or some antidote for poison, which today we use with safety. It is the same here. Properly, carefully and systematically developed, mediumship should present none of these dangers or difficulties, but should, on the contrary, bring the student into touch with higher planes of thought and activity and enable him to approach the more angelic sphere of being, as by careless or wrong development he will as surely come in contact with spirits of the opposite nature. For this reason I again urge the student to study and practice carefully and cautiously as he proceeds, so that none of these un‑ pleasant or terrifying experiences may at any time come to him.

CHAPTER XXIV

PRAYER, CONCENTRATION AND SILENCE

THE subjects treated in this chapter will probably be of special value to the student, after that immediately preceding it. We have already spoken of the value of prayer, in certain cases, and it may be said that both silence and concentration, under certain conditions, will always prove of great value,—not only in the cure of obsession but at all other times when dark clouds loom up on the horizon. In order to secure the best results from these psychological processes, however, they should be practised according to certain laws, and the reason for their operation should be thoroughly understood by the student.

WHAT "THE SILENCE" REALLY SIGNIFIES

What is called "the silence" in general New Thought philosophy, is a peculiar psychic state into which the student enters in order to secure certain results. As the term implies, quiet or silence are necessary factors, but they are mere means to an end; silence in itself would achieve nothing. It means that in this condition, thereby induced, certain practices may be followed, which produce the desired results.

Concentration is the focusing of the entire being at any given moment upon a central point of interest, either inward or outward, as the case may be. It may be upon some object, or some inner thought or psychic condition. Concentration upon objects is usually employed as a mere outward exercise to train the mind to act according to instructions, so to speak, so that when the time comes it may be employed in useful and helpful psychic activity.

THE VALUE OF CONCENTRATION

Concentration means power. The more we concentrate on anything the more certain it is to be accomplished and the better will the results be. Just as a number of streams meeting at a certain point will create a rushing, mighty current, so in the same manner will scattered psychic activity and forces, if brought to a common point, produce certain powerful results, which may be centred or turned into one direction or into another. One of the chief practical uses of concentration is the use it may be put to to hold or bind the self together. We should never let it be scattered in a variety of separate channels of expression, but rather concentrated into one single, powerful unit. Just as the strands of a rope may become separated, so the mind may become disintegrated and lose its initial power. In this weakened condition, it can be acted upon by other minds and forces, just as the single strands of rope can be broken; but the whole rope would resist any strain put upon it. The mind, when concentrated and acting under proper direction and control, is similarly impervious to all outside influences which may not be desired; and at the same time is itself a powerful factor either for good or evil.

CONCENTRATION EXERCISES

A few simple exercises in concentration may here be given which will be found useful not only in psychic development but in every phase of life.

1. Read a page of some heavy technical book, the meaning of which is not at all clear to you. Re-read the page with the determination to understand what it means. Read this over a number of times, if necessary, never letting your will relax for a moment, but determined to understand the thought of the author. If you do this, you will, after a few readings (the number varying) be enabled to grasp fully what is meant.

2. Place a watch in front of you, and look at the second-hand until it has completed the circle marking the minute. During this process never let your thoughts wander from the second-hand for a moment. Concentrate upon it fully. You will probably find that your thoughts are wandering, and that you cannot even for the space of a minute fix them absolutely under your control. Practise this until you have succeeded in accomplishing it.

3. Call up before your mind's eye a picture of a certain living friend. Hold this image in front of you as long as possible, making the details in every way as clear as you can, by endeavouring to fill out mentally the colour of the hair, of the eyes, the complexion and any peculiar markings that you can remember. Now when the face is vivid before your mind's eye, see whether you can discover any peculiarities in the face hitherto unknown to you. If you note anything of the kind, ascertain, the next time you see this friend, whether or not these impressions are correct.

4. Call up before your mental vision the face of some dead relative or friend. Concentrate upon it, holding it firmly before you in space. Study it closely, filling in all details as before. Finally, when you have held this vividly for a minute or so without wavering in your

attention, open your psychic or mental ears, so to speak, and see if you can receive any message from the person whose face is before you. This will be found a very useful practice, on occasion, at the beginning of your mediumship, and will enable you possibly to receive direct communications when you have tried in every other way and failed. You will not be able to do so, however, until you have mastered fully this faculty of concentration.

THE DYNAMIC POWER OF THE MIND

Having progressed so far, you may now concentrate upon certain mental or psychic processes, while "willing" or demanding that some return be made as a result of your volitional activity. Remember that every thought you send out into the universe attracts to itself others of a like nature, and ultimately returns to the sender with added power,—just as a boomerang returns to the thrower. Altruistic thoughts, such as "love, justice, forgiveness," etc., will therefore return to the owner, and make him happier for having thought them.

Evil and malevolent thoughts will, on the contrary, return and make the sender more unhappy and more innately evil in consequence. The path we travel, whether it be upward or downward, always gets easier as we proceed. We are helped along not only by the powers of good or darkness, but by our own thoughts and their results.

Thoughts are things! No thought is ever annihilated, and there is evidence to show that thought can take material form on occasion, and influence, either for good or evil, those at a distance. This will be more fully explained, however, in subsequent chapters.

THE VALUE OF PRAYER

We must now say a few words on *prayer*, and its great value to one who sends out the prayer thought. There are many who believe that prayer is "superstition" since they do not believe in a personal God, who grants or answers prayers, but rather in an impersonal, "Creative Power," which orders all things according to unvarying laws. Even on this theory, however, prayer, under certain conditions, is fully justified; for, in the first place, as we have just seen, helpful and wholesome thoughts tend to bring their own reward. In the next place, prayer is an auto-suggestion of great value, and its influence upon the mental and physical life is frequently very great. In the third place, prayer will help and buoy you up by giving you added confidence and belief in your powers. In the fourth place, inasmuch as telepathy is a fact in nature, you may, while in that condition, reach the minds of other human beings, who can help you and will actually do so, without knowing why. The many interesting cases which may be found, of "answers to prayers" (bringing a material return) fully justify its use from that point of view. Fifth, you can doubtless reach, by telepathy, friends in the spirit-life, who may be brought into more or less direct touch with you, during the prayerful condition of mind which is certainly closely akin to certain phases of subjective mediumship. They may in this manner be made aware of your condition for the first time, and will then endeavour to help you.

Sixth, by prayer you may bring yourself into harmony with the great Cosmic Currents of Good, which, as before explained, are playing hither and thither upon our Universe in much the same way as light, heat, gravitation, electricity and other material forces play or act upon it and us.

All these material factors must be taken into account, apart from the possibility that there is a receptive, loving and protecting power in the world, which is capable of helping us in time of need.

PRAYER IN OBSESSION

In obsession cases particularly, prayer is of value, because of the relief from tension and the wholesome mental attitude induced. Just as a drowning man will clutch at a straw, so those who are in terrible distress will frequently resort to this practice, when they would not think of doing so at any other time; and, in a sense, they are justified in so doing. There is an old saying that "Man's extremity is God's opportunity." It may be that prayer, in the ordinary sense of the word, is not needed during an ordinary healthy life, provided that it is lived in accordance with the laws of nature and according to its own highest mental and spiritual insight. At the same time, there may be occasions when it is justified and helpful, and certainly it has proved so in certain cases of difficulty and obsession.

HOW PRAYER CURES

The beneficial effects of prayer may be explained in many cases, quite simply. As explained in the chapter devoted to the Subconscious Mind, certain groups of thoughts tend to become bound together,—forming what is known as a "complex." If this activity be wholesome, the result is beneficial, and, in fact, all our educational processes depend upon this complex-formation. On the other hand, these groups of thoughts may be harmful, in which case they tend to press upon the mind from beneath, in much the same way that physical tumours might press upon some healthy structure in the body, and impede its functional activity.

The mind, therefore, becomes "diseased" by reason of this "pressure," and will only resume its wholesome attitude when this "pressure" is removed. By means of hypnotic suggestion and spiritual treatment, the mind may be opened up and explored, and this "complex" found and removed. This once done, the mind is restored to its wholesome activity, and the cure is complete. This is known in technical-language as the "Purging Treatment." As soon as the unwholesome load is removed, the mind is cured.

Now, in prayer, when a full and free confession is made, this same purging process occurs. The mind is freed from its burden and is consequently restored to health by its own inner nature. This being so, it may be seen that prayer, as such, is a real curative process and in many types of obsession and similar cases, it may be employed effectually, as before said, as a therapeutic measure of great value in curing the sick mind.

CHAPTER XXV

THE HUMAN "FLUID"

THE human body is charged with a certain magnetism which differs from all other forms of magnetism and electricity in the world. All other forces, of which we have knowledge, are *non-intelligent*, and have to be guided and directed by mind or by some law, in order to bring about any definite or desired result. It is therefore meaningless to explain the continued and consecutive movements of the planchette-board or any similar instrument by saying that it is due to "magnetism" or to "electricity" or to any similar power. These are all blind forces and must be directed, in order that any intelligent result may be obtained. The vital magnetism, which is present in the body, is also a blind force, but is under the control of the subconscious mind, and, under certain conditions, to be spoken of later, may be played upon or manipulated by external intelligences. In this way the various results are obtained.

NATURE AND PROPERTIES OF THE FLUID

This vital magnetism appears to be fluidic or semi-fluidic in form and capable of flowing from one organism into another. It is upon this principle that the various magnetic cures are based,—the fluid running from the operators' fingers into the body of the patient treated.

That this fluidic energy is present in any human body may be proved in a number of ways:—

In the first place, the human aura, which I described in an earlier chapter, is partly a manifestation of this vital activity, the colours being the varying vibratory counterparts of the energy radiated.

In Psychometry, again, it is this vital energy which passes into objects, impregnating them with its fluidic properties.

Each individual has his own peculiarly constituted and personal vital magnetism, and this differs from all others in quality and properties. A fully developed psychic is enabled to distinguish these, one from another, and a medium in trance may be enabled to get into communication with a deceased person through or by means of this fluidic impression left upon it, as explained in the chapter devoted to Trance. One or two practical examples or exercises may serve to show the student the reality of this fluidic emanation, and he may employ these to convince his sceptical friends also of its reality.

EXPERIMENT TO PROVE THE EXISTENCE OF THE FLUID

1. A very simple test is the following:—Hang a dead-black cloth over the back of a chair and see that no light falls directly on the cloth. The light in the room should be somewhat subdued and you should stand between it and the cloth, so as to throw your hands, held against the latter, into shadow. Now approach your two hands one to the other and touch the finger-tips together, the hands being otherwise opened wide, palms turned towards yourself and thumbs pointing toward the ceiling. In this condition you will probably find that, as the first and fourth finger-tips touch, the second and third fingers have to be bent considerably to touch one another. The hands should be at a distance of about three inches from

the black cloth and about 15 inches from your face.
Hold the finger-tips together for about 30 seconds; then
very gradually pull them apart and you will see, coming
from and joining your fingers, streams of whitish, misty
vapour, which is the fluid connection between the hands,
which you have established by the previous contact. If
you move the fingers slightly up and down, after they
have been separated an inch or so, you will find that the
streams or bands of light follow the fingers, still connecting them, which will prove that it is not due to hallucination or to what is called "persistence of vision."

HOW TO MAGNETIZE WATER

2. Place two glasses of water side by side on the table.
Over one of these place the tips of your fingers, held
together so as to form a point as much as possible. Hold
these over the water in one glass for four or five
minutes, *willing* that your vital magnetism should pass
into the water and *affect* it. If now you ask a sensitive
person, who has not seen you perform the experiment, to
pick out the glass of water which has been treated mag-
netically, he will be able to do so almost invariably, and
will tell you that the water sparkles as though charged
with some effervescent gas.

LIFE PRESERVED BY THE HUMAN FLUID

Some recent experiments, conducted by a group of
scientific men in Bordeaux, France, have proved con-
clusively that the human body radiates a form of vital
energy which may be extremely powerful in its results.
A lady, possessing the power of projecting or externalizing this vital energy in a remarkable degree, discovered that by placing her hands for fifteen or twenty

minutes daily for a period of two or three weeks over certain dead objects, such as oysters, canary-birds, fishes and even larger animals, such as guinea-pigs and rabbits, she was enabled to preserve them indefinitely; that is, instead of decomposing as they ordinarily would have done, they dried up or mummified and were preserved for months with no change whatever taking place in their substance! They never decomposed!

PROOFS OF THE HUMAN FLUID

This fact was fully endorsed by several chemists and physicians who studied her case and they stated that this was due to the fact that the vital emanation coming from her body killed or destroyed the bacteria usually present in all these bodies after death. This could be traced with the microscope. For example six oysters were allowed to decompose partially, while six were treated by her. The six that were treated never decomposed at all, but dried up or desiccated without any putrefaction. Now, when the other half dozen oysters had partially decomposed and bacteria could be seen under the microscope, Madam X. was requested to place her hands over the oysters and treat them. After fifteen minutes' treatment, they were again examined and it was found that thousands of bacteria had been killed. At the end of a few days they had all disappeared and the oysters dried up and thenceforward no decomposition whatever was noted!

This is a very striking proof of the reality of the human fluid and its peculiar action in certain cases. There is evidence to show, however, that in other instances its action is different from this, and that it imparts life and energy rather than proves destructive, as in the above unique case. Many persons have the power of preserving the life of flowers by treating them with their hands in a similar manner every day, and the student might well try this experiment and see to what extent he can preserve the life of certain flowers—others of a like nature being preserved at the same time by another person and under similar conditions to note the difference, if any, between the two sets.

It is this vital magnetism, which, projected beyond the bodily limits under the action of the will, is responsible for many physical phenomena, as we shall see in Chapter XXXVIII.

HOW MATERIAL OBJECTS BECOME CHARGED BY THE FLUID

Material objects, particularly of a sponge-like nature, such as wood, are capable of being charged up very highly by this vital magnetism, and when this is the case they come "en rapport" with the medium, who is enabled to move or manipulate them from a distance by his power of will, because of this vital, fluidic connection. We shall speak more fully of this, however, in the chapter devoted to physical phenomena.

It may be proved experimentally, also, that this fluidic magnetism is either capable of sensing pain or is the means whereby pain is carried from the nerve-centres to the consciousness.

EXTERIORIZATION OF "SENSIBILITY"

Under certain conditions the fluidic body, which is the inner part of the physical body and acts as its "double," may by hypnotic and magnetic processes be removed entirely from the physical body, in which case it may be acted upon by suggestion from others present at the

time. For example:—Colonel Albert De Rochas, of
Paris, succeeded in entirely disengaging or separating
the fluidic body of his subject from the physical body,
and gradually removed it to greater and greater distances until it stood *several feet* from the entranced
subject's physical organism! He then pricked the surface of the fluidic body with needles, and the sleeping
subject experienced these sensations of pain in her own
physical body at a spot or point exactly corresponding
to the part pricked on the etheric body.

"REPERCUSSION"

This seems to show that there is a direct vital or magnetic link between the etheric and the physical organ-
isms, and that injury done to the one re-acts upon the
other by means of what is known as "repercussion."
This is a very significant fact, when we remember that
in materializing séances, it sometimes happens that the
figure is seized or in some way injured by the sitter, and
the entranced medium is injured in exactly the same
way that the materialized figure is injured. This fact
has long been known to experienced spiritualists.

This curious fact has also great significance and throws
an interesting side-light on many of the phenomena of
so-called "witchcraft." We know that many of these
stories relate that the witch, assuming another form,
visited other scenes or localities, and if cut, shot or
injured there, she herself was found next day to have
received these exact injuries, though lying in her bed
at some distance from the scene of the event in question!

Such stories certainly appear more credible, when we
take into consideration the above facts, for both sets of
phenomena seem to depend upon "repercussion."

HOW THE HUMAN FLUID MAY IMPRESS
PHOTOGRAPHIC PLATES

The human fluid may also be proved to exist by means of photography. If a sensitive plate be wrapped in black paper and the hands of the psychic or medium of suitable temperament be placed upon it, the fluidic radiation coming from the hand and fingers will influence the plate through the enveloping black paper, and the impress of the hand will be found upon the plate. This can only be accounted for by supposing that the fingers became in some way radio-active during the experiment.

Many psychics can go further than this, and can impress upon the plate an image or figure of their thought at the time. Thus, when holding the plate between their palms or on their forehead or against the solar plexus, and thinking of a sheep, a cat, a chair, etc., the image of a sheep, cat or chair is impressed upon the plate. Experiments such as these may be tried by any student, and are of extreme interest and also of value, scientifically, when they are successful. It is to be hoped that many readers of this book will try experiments of this character and report any results they may obtain!

CHAPTER XXVI

SELF-PROJECTION

By "self-projection" is meant the faculty or ability to send out or cause to travel to a distance the etheric self or "double," by an effort of will. This seems to be, to some extent, inherent in some individuals, and occurs with them spontaneously and almost against their will. They go into trance and, at the end of a certain time, find that they have left their bodies and travelled to some distant scene! This, however, is rare: in the majority of cases the power has to be developed by long and assiduous cultivation.

OF WHAT THE "THOUGHT-BODY" IS COMPOSED

Before I come to speak of this projection of the self, a few words are necessary as to the nature and composition of this etheric body, or double, which is thus projected.

The physical body is composed of millions of tiny cells, and in each cell there is a centre or nucleus of energy. This centre is so infinitely small that it cannot be detected even by the highest-powered microscope. All that we know is, that physical matter in the cell is in some way vitalized or rendered living when it comes in contact with this vital centre. The source of the energy is invisible and cannot be determined by us. It seems to well-up from nowhere. Now, this centre of energy constitutes a sort of psychic point or cell of its own, and as there are millions of them in the body, corresponding to the number of physical cells, it is obvious that there are millions of vital cells which conform exactly to the shape of the body, since they correspond to its physical cells in life. These psychic centres have been called "psychomeres," and their bulk is estimated at about one millionth that of the physical body. The density of the etheric double, therefore, would be about one millionth as dense as the physical body. The combined weight of these psychomeres has been variously estimated, but probably varies between that of ten postage stamps and one ounce and a-half. This would represent the weight of the "astral" or etheric body, and is such that it would float slowly upward through the physical atmosphere, as would a balloon. This fact coincides with what we know of the gradual floating upward of the spiritual body after death.

INNER, FINER BODIES

It is this body which we inhabit after we discard the denser physical frame. It is not necessary to suppose that our consciousness is scattered throughout the *whole* of this body any more than it is at present. The centre of spiritual activity and the power of the will and mind may be a point of force, so to speak, within this etheric double and we may utilize and animate it just as we utilize and animate the physical body in this life. After a time it is probable that we discard this etheric body, to assume one of even lesser density, and that this process continues a number of times, until the spirit ultimately inhabits one of such infinitely fine matter,— if such it can be called,—that it is practically a mental or spiritual body. This is what we learn from many "spirits" who have communicated such facts to us. It is this body, therefore, which becomes disengaged from the physical body during life and goes on trips or excursions—carrying with it the consciousness of the individual and returns to animate the physical body at the end of a certain period of time.

POSSIBLE DANGERS AND HOW THEY ARE OVERCOME

When this disunion or severance takes place, there is always a connecting link, a "magnetic cord," which unites the physical and the etheric bodies. If this cord were to get broken, for any reason, re-animation would be impossible, and the death of the body would take place. This is the great danger attendant upon experiments of this character; but such a phenomenon is only possible in cases of very deep trance, where the separation is almost complete, and very little serves to disconnect it entirely. It is highly improbable that any but the most advanced student could reach this stage, and when he has reached it, certain mystical, inner practices may be resorted to, which would offset this possible danger.

It is this body which is occasionally photographed, and many so-called spirit-photographs are in reality photographs, not of discarnate but of incarnate spirits; that is,—they are wandering "doubles" of spirits still in the flesh.

Again, many apparitions and figures seen in haunted houses are of this nature. They constitute the projections of living persons rather than those who have passed over, and it takes an experienced psychic student to distinguish between the two types of figures. They have been known to appear at séances, also, in the form of "materializations."

THOUGHT FORMS: HOW THEY ARE BUILT-UP

In addition to these etheric bodies or "doubles," there may also be mental or thought-bodies, created entirely by the mind and will of the subject. Thus, in a case known to us, a clairvoyant was sent on a trip to the house of a friend and asked to describe the individual whom she found there. She described a certain person in detail—hair, eyes, features, etc., given at great length. When the psychic had finished and recovered full consciousness, she was told that her description was entirely wrong, and that no such person existed in the house in question, and that her description was throughout erroneous. In order to prove this, a journey was made at once to the house of the subject in question. When the facts were stated, he replied, that although he himself did not in any way resemble the clairvoyant description, this corresponded exactly and in minute detail to a character he was creating and writing about in his book! In other words: his thoughts had created the figure so vividly that it actually lived for the time being as an objective entity, and was seen as such by the entranced clairvoyant.

We can see from this, then, that "thoughts are things." They assume shape and, in a certain sense, live in the physical world. All our thoughts have a definite shape as well as a definite colour, and the more advanced students along the Path of Development can see and describe these thoughts, we are told, as clearly as we see objects.

HOW THE SELVES MAY BE PROJECTED

If this be true, it has a very significant bearing upon cases which occur and have been reported in the past.

For example, the reader will doubtless recall the case of "Dr. Jekyll and Mr. Hyde" by Robert Louis Stevenson (a most important case for all students to study). Here, as we know, the original individual finally became two. Dr. Jekyll developed another self, calling himself Mr. Hyde. Dr. Jekyll was kindly, helpful and sympathetic; Hyde was evil, malicious and wholly repulsive. These two selves were developed in the original person, and the split between them became greater and greater as the months went by. Finally Mr. Hyde assumed complete supremacy and Dr. Jekyll vanished for ever! In this case it was not a mere change of personality which could be accounted for on any psychological grounds; it was an actual physical transformation. Apparently there were two separate bodies, which were transformed one into the other!

Suppose now, that the good self, Dr. Jekyll, also the bad self, Mr. Hyde, existing as separate, mental beings, each had the power of self-projection. They would each create, by their own thoughts, a separate body, and this being would resemble, in outward appearance, the thoughts which created it. Cases of this character might, therefore, exist, and might conceivably be explained on scientific principles.

We must be careful, then, of the character of the thought-self which we build up, for if this resembles outwardly its inner structure, we may (many of us) come to resemble monstrosities rather than human beings, at some stage of our development, when the plane is reached where thoughts predominate and shape the expression of the self! (This idea has been graphically portrayed in John Uri Lloyd's book, "Etidorhpa.")

PRACTICAL INSTRUCTIONS FOR SELF-PROJECTION

This inner, etheric body, which is expelled to a distance by the power of will, in cases of self-projection, may be released and projected by the student after a certain amount of practice. He should go about this cautiously, feeling his way, as it were, but proceeding more or less along the following lines:

Place yourself in a perfectly composed attitude, either on a couch or in a large chair. Close the eyes and breathe deeply for a few minutes, all the time holding the mind on a central point of concentration. Travel over your body in thought, and at each point or spot dwelt upon by you, *will* that your etheric body becomes detached or loosened from its connection with the physical body. As you begin to gain control of this process, you may hear or rather "sense" a process of separation taking place, resembling a "click," and inwardly feeling like the disconnection of an electric current. When this has been completed at one point, travel to another. Do not try too many on any one occasion, and always be sure to restore by an effort of will the original connected condition before you terminate the experiment.

FURTHER DIRECTIONS AND ADVICE

After you have gone round your body in this way, and have succeeded in disconnecting it more or less completely, you should then call up before you, in space, a certain distant locality, such as the room of a friend, and, throwing the whole force of your being into a single determined effort of will, force yourself mentally to leave your body and travel to the locality before you. If you feel that you are losing consciousness, or

that everything is "going black" before you, discontinue
the experiment at once and return to your physical body.
If you can keep your self-consciousness active, you may
safely travel to any distance,—feeling assured that you
will be able to return whenever you want to and reanimate your own physical frame. All this, of course,
takes time and persistence of development, and cannot
be acquired in a few days. Moreover, I would advise
the student not to attempt this process, until he has
progressed further in his studies and read the advice
contained in the last chapter.[1]

Should he, however, make up his mind to do so, he
should proceed along the above lines, advancing cautiously all the time and never allowing himself to lose
consciousness at any stage of the proceedings.

When he has acquired this power, he will have in his
possession a wonderful knowledge, and a means of acquiring information and spiritual insight which others, who
have not developed it, are totally unable to comprehend.

[1] Further advice is to be found in my book "Modern Psychical
Phenomena," in which a chapter is devoted to this question.

time of the individual. But how about those which occur *after* death? Here we should have to assume that some other process was involved, or else extend our belief so as to cover and embrace the action of discarnate spirits.

PHANTASMS OF THE DEAD

One theory of these apparitions (seen after the death of the person they represent) is that they embody the thought of the dead person. For example: an individual spirit may continue to think over its life and the scenes of its varied activities, and these recollections and thoughts, influencing the minds of those still living, by means of telepathy, would cause them to see the phantasmal image of the person thinking the thoughts. This, however, is a question which we shall discuss more fully in the next chapter. For the present it may be said that this is one theory advanced to explain so-called phantasms of the dead, or "ghosts," as opposed to phantasms of the living, and phantasms of the dying.

GHOSTS THAT TOUCH!

There are many cases of apparitions, however, which cannot be thus easily explained by assuming that they are the projection or telepathic influence of a living mind, or the mind of a discarnate spirit. In many cases, they seem to be real substantial beings,—to occupy space and exist as real semi-solid, or material phantoms. Those who have been convinced of the reality of an etheric or spiritual body, need have no difficulty in assuming that it is this body which is seen at such times, and in many cases we find strong evidence for supposing that a body of this character actually exists. For example: In one historic instance, a doctor and his wife

both saw the figure of a woman standing at the foot of
their bed, and saw it cross the room and place its fingers
over a small night-light, which was burning on the
mantelpiece. At the moment the phantom thus placed
his hands over the light, it was extinguished and the
room was left in darkness!

Here it is difficult to suppose that any thought-creation
or "telepathic hallucination" of any character existed,
for the reason that a physical phenomenon was produced
and no hallucination could have done this.

MATERIALIZED PHANTOMS

Again, in many cases, the phantasmal form or apparition is seen to open doors, lift curtains, raise bed-clothes,
etc., and in such cases, again, we must assume that a
real phantom exists. The problem is thus more complicated than at first appears, and, as Mr. Andrew Lang
remarked, "Consequently, if these stories are true, some
apparitions are ghosts,—real objective entities filling
space. Hallucinations cannot draw curtains, or open
doors, or cause thumps—not real thumps—hallucinatory
thumps are different."

Dr. Burns tells of a gentleman, who, in a dream,
pushed against a door in a distant house, so that those
in the room were scarcely able to resist the pressure!
Now, if this rather staggering anecdote be true, the
spirit of a living man, being able to affect matter, is
also, so to speak, material and is an actual entity, an
astral body. These arguments then make in favour of
the old-fashioned theory of ghosts and wraiths, as things
objectively existing, rather than the view that *all* these
"ghosts" are necessarily subjective in origin.

PHANTASMS CREATED BY THOUGHT

These phantasms are doubtless thought-bodies, in many cases constructed by the operating intelligence itself. One interesting fact in this connection is this: that it is nearly always stated by those who have seen figures of this kind that the phantom is clear and plainly visible about the head and the upper part of the body, but that the apparition dwindles down to a vaporous film toward the feet. In other words, the upper part of the body is much clearer than the lower part.

If the phantom were a definite thought-creation, this is only what we should expect. For we think of the upper portion of our bodies much more than the lower portion; we are more *conscious* of our head and shoulders, and the upper portion of the trunk and the hands and arms, and only vaguely conscious of the legs and lower portions of the body. This is exactly what we find in apparitions; and it would therefore seem that the figures are clear in outline just to the extent that the operating intelligence is intensely conscious of the appearance of the body he is creating or building up.

PHANTOMS WHICH IMPART INFORMATION

There are also certain cases on record in which the phantom has given the recipient of the experience some important information, which he did not know previously —where certain papers are hidden, etc. Such cases certainly prove that an independent intelligence is there— a spirit which is thus manifesting its presence. It must be admitted, however, that most apparitions are purposeless and meaningless; but this is easily accounted for by supposing that we see, at such times, not the spirit itself, but its mere projected thought—a phantom created by the spirit, rather than the spirit itself. Most apparitions are, doubtless, of this nature.

We have seen that there are apparitions of the living, of the dying and of the dead—mostly attached to human beings. When they are attached to *localities* they become local phantasms, or cases of "haunting," of which we shall speak in the next chapter.

EXPERIMENTAL APPARITIONS

In addition to these, there are so-called cases of "experimental apparitions," in which an individual succeeds in creating a phantasmal figure at a distance, by an effort of will or thought. These closely resemble certain cases of self-projection on the one hand and cases of witchcraft on the other, and form an intermediary between them,—since on the one hand they are mere mental pictures and on the other they are real physical entities. Experimental apparitions, then, seem to bridge the gulf between these two types of phenomena, and form a connecting link.

Apparitions may be induced experimentally by willing very strongly, just as you are falling asleep that you will appear to a certain person at a certain time, and, if this is properly managed, it will be successful in a large number of cases. This may also be induced experimentally by means of hypnotic suggestion or magnetic or mesmeric processes, and, when in the trance, the spirit of the sleeper may be directed to a certain locality and there seen by those present. The natives of West-Africa claim to be able to do this more or less at will. They can project the "double" or "etheric body" and, so to speak, materialize it at the other end!

HOW TO CREATE THOUGHT-FORMS

The same laws which prevail in many of the previous exercises also rule here.

The student should see to it that he retains a grasp of his own personality and does not lose control of his inner self at any stage of the proceedings. As he progresses in his development along these lines, he should endeavour to make the apparition which appears "at the other end of the line," so to speak, more or less solid. After he has once succeeded in the process of projection, he should throw all his will into the effort to make the projected form more and more substantial, and to will that his self-consciousness and activity be actually transferred to the distant scene. In this way he is not only seen by others, who may happen to be present, but is also enabled to see for himself what is actually going on in that place, and obtains, at the same time, a clairvoyant vision of the surroundings in which he has appeared. In this way both the psychic and those who perceive the created figure mutually exchange experiences; and this process should be continued until the projected double becomes so solid in structure that it cannot be distinguished from a real physical being. There are many advanced psychic students who claim that they can actually create and project to great distances material bodies of this nature.

CHAPTER XXVIII

HAUNTED HOUSES

As explained in the last chapter, when apparitions become fixed or attached to one locality, they constitute what is called a "local haunting," and the place they influence is commonly called a "haunted house." This is the ordinary or common theory of haunted houses, and the average person probably assumes that the figures seen in such houses are material, and the picture he forms of the ghost is that it is a sheeted figure walking about, up and downstairs, and clanking chains after it! There are probably few, if any, psychic students who believe that houses are haunted by figures of this description, and opposed to this view is that of ordinary science, which contends that there are no haunted houses at all —the figures seen within them being merely the product of expectancy, suggestion and excited imagination!

THE EXPLANATION OF HAUNTED HOUSES

All those who have carefully investigated the subject, however, come to the conclusion, sooner or later, that there *are* genuine haunted houses. The question is: what constitutes the haunting and how are such cases to be explained? Many psychic students have specialized, so to speak, in this subject of haunted houses, and have formulated various theories to explain cases of this character. The following are the most important theories which have been advanced:

1. That one person or group of persons, forming a

243

family, have experienced certain psychic phenomena in
the house in question, and these formed the nucleus round
which gathered impressions, noises and psychic experiences of all kinds. From a small beginning great results
sprang, elaborated by their own minds. Now, when
these people moved away from the house in question,
and other tenants occupied it, this second group was
influenced by the thoughts, emotions and impressions of
those who had moved away, so that they in their turn
began to see signs and hear strange sounds,—inquiry
revealing the fact that the house had the reputation of
being "haunted," and their own imaginations would
magnify the significance of all they had seen or heard.
In other words, this theory contends that telepathy or
influence from living minds is the all-sufficient explanation and alone serves to account for the facts.

TELEPATHY AND "PSYCHIC ATMOSPHERE"

2. The second theory advanced is that telepathy from
the dead is the true explanation—the phantoms seen,
etc., being produced by the influence of minds of deceased persons. On this theory the figures and phantoms
are not objective or real any more than in the first case,
but are telepathic hallucinations, just as truly, though
they have an objective basis of reality, inasmuch as
they have originated in the mind of a deceased person.
Dreams or thoughts of the dead constitute, therefore,
the basic principle of explanation on this theory.

3. The next theory which is advanced is that some
subtle psychic atmosphere permeates the walls of the
house in question, and that this atmosphere influences
or impresses all those who live within it. There is
much to say in favour of such a theory, and the pre-

vious chapters on the Aura, Psychometry, The Human
Fluid, etc., will lend a certain amount of support to it.
At the same time it is difficult to see how a general and
impersonal atmosphere of this character could translate
itself into definite figures or forms, particularly when
these speak and convey information unknown to the
seer. I shall say more of this later.

ASTRAL BODIES AND THOUGHT-FORMS

4. The fourth theory to be advanced is that the figures
seen are the astral or etheric bodies of spirits who return
and constitute the haunting,—being present in actual
fact. This is the nearest approach to the commonly-held
theory of the figures seen in haunted houses.

5. The fifth theory is that such figures are thought-
forms, created by some distant, living or discarnate mind,
and projected into the house in question, where they
assume more or less definite and tangible form. This
is, in a sense, a process of self-projection, but the phantasm is always seen in a certain place as though magnet-
ically drawn to that locality.

THE NATURE OF THE FIGURES SEEN

Which of these theories is the correct one? In my
own estimation there is much truth in all of them, and
no two cases of haunted houses are due to the same
cause or depend upon the same conditions. All five of
these causes may be operating at the same time in any
one house, or any two, three or four of them may be.
Indeed, to judge from the complex nature of the phenomena seen, it is highly probable that such is the case.
There is strong evidence, in fact, to make us believe that
the ordinary hallucination theories will not serve to ex-

plain the facts. For example: these phantasms often produce physical phenomena, as before explained,—such as opening doors, lifting curtains, snuffing candles, etc. Mental images or pictures could not do this. Again, animals often see, or appear to see, apparitions in haunted houses, and show all the signs of fear, such as trembling, sweating, etc.

In the third place, figures are often described differently by different individuals. For example: A. would describe a full-face view of the figure, while B. would describe the figure in profile. If a real figure were standing where both percipients saw it, this description would be correct. Such cases certainly tend to suggest that a real figure, and no mere hallucination, was present.

In the fourth place, apparitions have been seen by two, three or more persons at once. These "collective hallucinations," as they are called, strongly suggest an external phantom and no mere mental picture.

PROOFS OF REALITY

In the fifth place, apparitions which have appeared to strangers occupying haunted houses have afterwards been identified on being shown the photograph of the person. For example: a gentleman sleeping in a house, reputed to be haunted, sees a certain figure, bending over him when he awakes at midnight. He notes details of dress, feature, etc., and also notes that he has never seen this person before in his life. The next day he is shown twenty photographs. From among the twenty he selects one as being the phantom seen in the house. The owner of the house then tells him that this is the person said to haunt the locality in question! Again we are driven to believe that more than mere hallucination is at work.

In the sixth place, these figures, seen in haunted houses, have occasionally been photographed, and this objective and physical proof of their reality is strong evidence that they are more than mental products.

Seventh: figures seen in haunted houses often convey, to the seer, definite information or give messages which the individual in question could not have known. This strongly indicates not only the reality of the apparition, but the fact that it is a discarnate spirit.

For these reasons, therefore, we must assume that haunted houses are actual realities, and that the figures seen therein are, at times at least, outstanding entities and represent more or less directly the individual they appear to portray.

SÉANCES IN HAUNTED HOUSES

Psychic students can test their power and at the same time conduct many interesting and valuable experiments in haunted houses. In an atmosphere of this sort, which is more highly charged with magnetism than the ordinary séance-room, psychic powers of any character should be quickly augmented and increased, so that messages could be obtained by speech, vision, automatic writing, crystal vision, etc. Whenever you hear of a case of a haunted house, therefore, you should make it a point to visit this house at once. It is not necessary to sleep in it a night, as many suppose, in order to test its character. Hold a séance in that house in the evening, and striking phenomena will probably result. Or, if you cannot gather together a group of interested students, sit by yourself and see whether you cannot obtain direct messages from the intelligences present. Experiments in automatic-writing, crystal-gazing, etc., may also be tried.

CLAIRVOYANT DIAGNOSIS OF HAUNTED HOUSES

Clairvoyants may also render useful service by visiting clairvoyantly haunted houses and ascertaining and describing, if possible, the source and cause of the haunting. Visit the house by means of a clairvoyant excursion, either spontaneously or when in a mesmeric trance, etc., and use your psychic powers to the utmost, to discover what you can regarding this house. When you find yourself inside it, look about and see whether you can "sense" any spirits, evil or otherwise, lurking within its atmosphere. Endeavour to sense the psychic atmosphere of the house and test the aura of those living within it. All houses, reputed to be haunted, may not necessarily be so, but the individuals themselves may be unbalanced or obsessed for some reason,—in which case the house itself would be free except from those influences which were drawn to it by the individuals residing therein.

Many persons, living in haunted houses, wish to be free from the depressing influences which sometimes hang about houses of this character, yet do not know how to proceed in order to rid themselves of these haunting presences. This is a very complicated question, and one to which psychic students have in the past given far too little attention. In my book "The Coming Science" there is a chapter entitled "Haunted Houses and their Cure" and I would refer all those interested to the work in question. An interesting case is there given of a haunted house which was "cured," so to say, by the following means:

HOW TO "CURE" HAUNTED HOUSES

A trance-medium, Georgia Gladys Cooley, was called in to investigate and do what she could, and, when in the house, went spontaneously into trance. In that condition her guides spoke through her and described the haunting "spirits." They were then charged to remove them, if possible, and undertook to do so. This they did in a somewhat striking and dramatic manner; and ended by reporting the fact that the haunting presences had been finally completely removed!

This is a very instructive case, and shows us that trance-mediums and their guides can be of very great service, in many cases, where the haunting assumes an unpleasant or evil character. Thus the nature of the haunting may be diagnosed clairvoyantly, and the cure effected through some trance-medium, and by the spirits who operate through him.

In some cases, however, the haunting may be cured by more simple means,—such as suggestion, lessening the psychic sensitiveness of those living in the house —by diverting the thoughts, by plenty of outdoor, physical exercise, toughening the aura, etc.

On the other hand, there are cases on record where haunted houses have withstood all attempts to cure them, and the inhabitants have ultimately been forced to move. Happily, cases of this character are rare. At all events, haunted houses present a fascinating and useful field, in which the psychic student can test his clairvoyance, or other psychic power, to advantage.

CHAPTER XXIX

THE DIFFICULTIES OF COMMUNICATION

THE process of communication is doubtless far more difficult and complicated than the average person believes,
and is even more complicated than most spiritualists
believe. As stated in a previous chapter, one of the
great objections to the reality of spiritualism urged in
the past, is that, if true, many more persons must communicate than now appear to do so, and that of the
thousands who die, more *must* come back than the few
who return through mediums! It was there pointed
out that the reason for this consists partly in the fact
that "good communicators" are comparatively rare,
and that there is necessarily a peculiar psychical condition which enables them to communicate through
mediums. In addition to this a medium must be present
at the time that an effort to communicate is being
made, and in many cases the recipient of the message
must also be reaching out to receive it, before it can
be given satisfactorily. In other words, the sender and
receiver of the message must stretch out their "mental
arms," so to say, at the same moment, before they can
shake hands across the Great Gulf; and if only one
does so, he fails to reach the one on the other side.

FACTORS AFFECTING COMMUNICATION

It has frequently been pointed out by scientific investigators of spiritualism that only after the reality of
the facts has been proved, does their detailed study

begin. For example: Supposing that a spirit can write through an entranced medium—she giving the messages in automatic writing. The fact once admitted, the scientific study of the case will only have begun, and such questions as these would then have to be answered: To what extent is the medium's spirit disconnected from the body while the communication is taking place? What is the degree of mental activity of the medium's spirit during the communication? Does the communicating intelligence act directly on the brain and nervous centres of the medium, or in a more roundabout manner, and if the former, upon what brain-centres does the intelligence act, and how?

If a communicator was in life a good visualizer or had a good memory, etc., would these factors assist him in the process of communication,—and if so, how?

These, and many similar questions, would have to be answered, and it is upon questions such as these that many psychical researchers have bent their energies for some time past. It is probable that several hundred years will have to elapse before these questions can be answered fully and the facts explained in detail.

DIFFICULTIES OF THE COMMUNICATING SPIRIT

Let us enumerate some of the difficulties which a communicating spirit probably has to contend with, in sending messages through mediums to the living. There is much evidence to show that the process of communication is a very difficult one, for, as soon as a spirit gets in contact with a medium and begins to transmit messages, he becomes more or less exhausted and suffocated, so to speak, by the dense aura or atmosphere with which he is called upon to come into contact. In many instances we read that spirits have to go away several times during the course of a séance, to revive themselves, and afterwards return, refreshed and clear-brained, to continue the communications. They experience great difficulty in holding their thoughts together, connectedly, during the process of communication. This does not mean that they are ordinarily in this confused state, but (very often) as soon as they come into contact with the medium's psychic atmosphere and magnetism, they become confused and their minds tend to wander as they would in delirium or in a state of trance. It is because of this that many of the messages we receive commence well but afterwards dwindle off into incoherence and triviality.

WHY MANY MESSAGES ARE "TRIVIAL"

This question of "triviality," however, is often misunderstood. The objection is raised, that spirits, if they really communicate, would tell us something more important than they usually do. As a matter of fact, however, this is only true in a certain sense. The ordinary social conversation between "spirits *in* the flesh" is not as serious as it might be, and it has been shown by actual experiment that human beings, when called upon to prove their own identity to another, *do* deliberately choose trivial incidents by means of which to identify themselves.

Another point is that trivial incidents serve best to prove identity, as some great philosophical discourse might be given by any intelligence, either in or out of the body, and would prove nothing to one longing to hear from his own dear one. In such a case personal, detailed and, so to say, trivial messages are often the most striking and the most convincing. The very triviality of many messages received through mediums is, therefore, their strongest point and not their weakest.

In addition to this there are, as we know, innumerable books written by spirits containing philosophical, scientific and religious truths of great value and importance.

INFLUENCE OF THE MEDIUM'S ORGANISM

Another reason why communication is doubtless difficult is that the communicating spirit is unused to the bodily organism of the medium. All of us have certain mental and physical habits which we form, and it is easier for us to do certain things in certain ways after we have done them in that manner a few times. If you were suddenly transplanted into the body of *another* person (say one of the opposite sex), you would find great difficulty in manipulating that body, so as to extract from it the best results—to think clearly and to speak and write clearly, when expressing your thoughts. It is precisely this difficulty which the communicating intelligence experiences in trying to communicate with us through unfamiliar bodies. Many of the habits and "tricks," so to say, of the medium creep into the messages, which are consequently often more or less similar to the language employed by the medium. This proves only that the spirit has to employ the medium's mental and bodily habits as best it can, during the process of communication, and that it is not as easy and concise as many persons imagine.

SYMBOLISM NECESSARY

Another difficulty presented is that the conditions on the "other side" are doubtless so different from any which exist here that they have to be explained in roundabout and symbolic language. If you had to explain colour to a blind man, you would find great difficulty in doing so. If you had to explain the feelings experienced while giving psychometric tests to one who had never experienced them, you would also find considerable difficulty. It is much the same in this case. There are no immediate analogies which can be drawn, and the result is that symbolism and a language which appears to us vague and unsatisfactory is often employed in describing the other side of life and the conditions which prevail therein.

DIFFICULTIES OF NAMES AND DATES

Names and dates furnish great difficulty for returning spirits. Dates, because of the fact that time is not recognized by them in the same way that it is with us. Names, for the reason that they do not represent concrete pictures or meanings, but are as a rule only a combination of letters having a certain sound. The word "chair" calls up to the mind a certain picture which can be visualized. On the other hand, the name "Robinson" calls up no such picture, except perhaps the memory image of some friend of yours by that name. If that memory-picture is revived in the communicator's mind, the medium can see this and describe it,—which is precisely what he does; but the name "Robinson" cannot be presented in picture-form (the most common form of representation) and consequently is not easily communicated.

As explained in the chapter on Dreams, our hearing-centres are less developed than our sight-centres, and for this reason verbal messages are less easily given and received than pictured or visualized messages. The difficulty in receiving names is explained largely because of this fact.

COMMUNICATIONS IMMEDIATELY AFTER DEATH

For some days after death, these difficulties are particularly great, and especially in the case of suicides. Dr. Hodgson, in his Report of the case of Mrs. Piper, says:

"That persons, just deceased, should be extremely confused and unable to communicate directly or even at all, seems perfectly natural after the shock and wrench of death. Thus, in one case, the spirit was unable to write the second day after his death. In another case, a friend of mine, whom I may call D., wrote,—with what appeared to be much difficulty, his name and the words:—'I am all right, Adieu,' within two or three days after his death. In another case, F., he was unable to write on the morning after his death. A few days later, when a stranger was present with me for a sitting, he wrote two or three sentences, saying: 'I am too weak to articulate clearly,' and not many days later he wrote fairly well and quite accurately, dictating also to Mme. Eliza, the amanuensis, an account of his feelings when finding himself amid new surroundings. Both D. and F. became very clear in a short time. D. communicated frequently, later on, both by writing and speech."

OTHER DIFFICULTIES

Other difficulties remain,—such as the probable inability of the communicating spirit to see the material world as we see it, especially at the time of communication, the difficulty of holding the mind together while communicating, the difficulty of manipulating the medium's organism, and the intra-cosmic difficulties, which exist between this world and the next.

Because of all these hindrances and impediments, spirits find great difficulty in direct communication and because of these facts, messages are, comparatively speaking, few, and in so many cases inconclusive. When a good medium, a good communicator and a sympathetic sitter get together, however, very striking and convincing results are obtained, as we know from the history of Spiritualism!

CHAPTER XXX

HYPNOTISM AND MESMERISM

THE word "Mesmerism" is derived from Antoine Mesmer, who founded the system and who performed all the early experiments in this field. It was known as Mesmerism for about fifty years, until an English physician by the name of Dr. James Braid coined a new word, "Hypnotism," from the Greek "Hypnos"—sleep, and this is the word which has been used almost exclusively from that date to this.

THE DIFFERENCE BETWEEN HYPNOTISM AND MESMERISM

The majority of persons would claim, at the present day, that hypnotism and mesmerism are identical, there being no difference between them. They are both due—it is said—to suggestion and the influence of the mind over the body. Very similar phenomena occur in both cases, it is true, but I believe that there is a difference between the two processes and conditions.

Mesmerism is based on the belief that there is a definite physical emanation or vital fluid, which passes from the operator into the subject while the mesmeric passes are being made over the latter's body.

Hypnotism, on the other hand, is due entirely to "suggestion"—the influence of the subconscious mind upon the body. There is no physical influence or effluence in hypnotic practice, and it is claimed that all the phenomena of mesmerism, apparently showing such influence, are in reality entirely due to suggestion.

As before stated, however, we believe that there is a difference between the two processes, and that hypnotism is due solely to psychical causes, but that, in mesmerism, the human fluid before spoken of, plays a part. As proof of this, I may cite (among other proofs) the fact that clairvoyance and many of the so-called higher phenomena are frequently obtained in mesmeric trance, while they are extremely rare in hypnotic trance. Other phenomena could be mentioned, but this will suffice for the present.

PASSES AND SUGGESTION

Mesmerism being due to the passage of a vital fluid from the body of the operator into the subject; contact and passes are essential. If, therefore, you wish to mesmerize your subject, you should make passes over his head, forehead, eyes and down the front of the body. All *downward* passes are sleep-passes and all *upward* passes are waking-passes. Placing the hands on certain nerve-centres of the forehead, and particularly between the eyes and over the temples, will help to induce sleep; also clasping the patient's hands and placing the point of your thumb in contact with the point of his thumb establishes the current and serves to induce the mesmeric trance.

In Hypnotism, on the other hand, passes are not essential, though they often help. In hypnotizing a subject it is common to ask him first of all to gaze at a bright object until the eyes tire,—when the lids are closed, suggestions of sleep are given, or the subject may open and close the eyes a number of times as you count, and this will serve to induce the initial stages of hypnotic trance; the deeper stages are induced by means of suggestion.

REMARKABLE HYPNOTIC PHENOMENA

Post-hypnotic suggestion is a form of treatment often resorted to and is a good subject for experimentation. It means that the subject performs, after awakening from trance, certain actions suggested to him when entranced. He remembers nothing of the suggestions but carries them out to the letter.

Many hypnotic subjects have extraordinary ability in calculating time, and can guess to a second the length of time which has elapsed between certain intervals or carry out "post-hypnotically" a suggestion given them in trance,—days or even weeks before.

Hypnotism is a useful method of opening up and exploring the subconscious mind. We are enabled to "tap" it, as it were, and get in touch with hidden portions of our being which we could otherwise never reach. Dreams may be analysed in this manner; also unpleasant thoughts, impressions, emotions, etc., removed and frequently undesirable influences banished by hypnotic suggestion. Hypnotism seems to reach a deeper stratum of our mind than ordinary waking suggestion, and because of this fact it is at times so useful. For instance the drink-habit has often been cured by hypnotic suggestion.

THE POWER OF HYPNOTIC SUGGESTION

Hence we see that there must be more in the hypnotic command than mere advice or persuasion, because thousands of drunkards have been advised not to drink, but they continue to do so, nevertheless! By means of hypnotism we are enabled to reach a portion of the mind so deep that it controls the whole being, and the result is that these deep-rooted habits may at times be removed and eradicated.

This is one of the distinguishing marks of the hypnotic state—that a more fundamental control over the body and mind is obtained, and, by reason of this fact, many cures of diseased conditions and abnormal states of mind have been recorded,—which have been otherwise treated ineffectually.

There is a difference between the hypnotic and the mediumistic trance, though not so great as that existing between the latter and the mesmeric state. In both the mediumistic and the mesmeric trance a form of "magnetism" is doubtless employed, and this connects them in a subtle bond of union. It is because of this that telepathy, clairvoyance, etc., are so often obtained in the mesmeric trance, which is closely akin to the condition secured by mediums, in which they obtain genuine mediumistic messages.

THE FEAR OF BEING HYPNOTIZED

Many persons are afraid of being hypnotized,—this fear being based partly upon valid reasons and partly upon superstition. *Properly induced by an expert*, the hypnotic trance is not injurious; on the contrary, it is often extremely beneficial, and, as before pointed out, quickens the mental and physical powers, removes bad habits, effects cures, etc. On the other hand, when hypnotism is applied by an ignorant or bungling operator, who does not know his business, the result may be very detrimental to the health of the person hypnotized. A state may be induced which neither the operator nor anybody else fully understands, for no one at the present time fully comprehends the nature of the condition thereby induced. The conscious mind is removed from its supremacy, and this is often a fatal mistake—particularly when there are evil influences at work, either within or without the subject.

If the operator is a sympathetic, careful, and qualified expert, mesmerism may prove highly beneficial, for evil influences may thereby be removed, by counteracting them and infusing into the subject a supply of beneficial "animal-magnetism" which is opposed to that supplied from opposite sources.

HYPNOTIC INFLUENCE FROM OTHER MINDS

Andrew Jackson Davis began his career as a medium by being mesmerized, and others could doubtless develop their mediumistic faculties in the same way; but one must be extremely careful in such a case to select a thoroughly competent operator,—one in whom he has complete faith, otherwise more harm than good may result. If you find that any one is trying to influence you against your will, you may overcome this by counter-suggestion given to yourself from within. If the person be absent, this may be purely imaginary on your part, and the operator in question may be entirely ignorant of the effect he is producing in you! There are thousands of persons in insane asylums all over the world who suffer from the belief that they are being "persecuted" by others at a distance, and that these others are endeavouring to influence them by hypnotism, etc. As a matter of fact nothing of the sort is the case, and their condition is purely the result of imaginary belief. Be most careful, therefore, that you fully ascertain and prove to your satisfaction the existence of this foreign influence before you take any steps

to offset it or even seriously believe that such influence is
being directed toward you.

HOW TO OVERCOME SUCH INFLUENCES

When once you have become satisfied that influences
of this character are being directed toward you, take
immediate steps to protect yourself—such as those outlined in Chapter XXIII, "Obsession and Insanity." If
promptly applied, this will effectually offset such conditions coming from outside minds. If you are in the presence of a person whom you feel to be influencing you, it
would then be best to take the precautions and steps outlined in the next chapter, devoted to "Personal Magnetism." This will prevent your passing under the influence of such a person. You need never fear that hypnotic sleep, even if induced, will last a great length of time
and that the subject cannot be awakened therefrom.
Sleeps of this character always terminate spontaneously
if they are let alone,—though it is always best to see that
a hypnotic subject is thoroughly awakened before he
leaves the care and supervision of the operator, otherwise
he may go about in a somewhat dazed condition for a
time, and may not be altogether responsible for his actions.

AN IMPORTANT WARNING

Somnambulism is a variation of hypnotic sleep where
the subject spontaneously performs a number of complicated actions and the subconscious muscular activities play a large part. A person who is subject to
somnambulistic attacks should *never* under any circumstances be awakened suddenly. It is a good plan to
speak to such a person and suggest to him, as to one in

hypnotic trance, that he return to bed; and, this done, suggest to him that it is impossible for such a condition to again occur, etc. Somnambulistic attacks of this character may often be cured by hypnotic treatment and properly directed suggestion.

PREVENTION OF HYPNOTIC INFLUENCE

An operator may prevent his subject from being hypnotized by any other person through forceful suggestions to his subject that he will be enabled to resist suggestions from any other operator—that he will have no effect on him, etc.

If you do not wish to be hypnotized at all, you may give similar suggestions to yourself. These Self-suggestions are called "Auto-Suggestions." Lightly given and persistently repeated, they will effectually prevent you from being influenced by any other person.

CHAPTER XXXI

PERSONAL MAGNETISM

WE all know the difference between a positive and negative personality; between an individual who is naturally successful and one who is not. The former seems to attract to himself success, happiness and prosperity; the latter seems to repel it. It is not necessary for a naturally positive person to say anything or to perform any action in order to make us feel this power within him. It seems to radiate silently from him as a form of power. Many times, doubtless, we have all stepped into a room, an elevator, etc., and immediately felt the strong personality and presence of an individual of this character, possessing much natural magnetism. They may know nothing of this power,—perhaps hardly realize that they possess it, although they do,—in many cases, to a remarkable degree. Properly developed and utilized, this power helped to make the great names in history. We may, all of us, cultivate and develop this power to a great extent by proper practice, and the degree to which we can develop it will make us successful accordingly, not only in the material things of this world, but will also enable us to achieve mental and spiritual heights which the ordinary person cannot attain.

"THE INEXHAUSTIBLE SUPPLY"

We must constantly bear in mind that there is an unlimited supply of Cosmic Energy, and this will develop personal magnetism to the degree to which we can draw upon it. Exercises for doing so have been given in a previous chapter. We must have confidence in ourselves and in our own powers; for "Confidence in self breeds confidence in others, and fear weakens both the brain that plans and the hand that executes." We must use suggestion rightly in our conversation with others, and, without appearing to do so, constantly give such suggestions as are likely to take root in the mind; and this must be hammered in by constant repetition. Finally, we must not waste the magnetism we may possess by nervous habits,—such as tapping on the floor or table with the fingers, pacing up and down the room, etc.,—in short, all unnecessary gestures. If we save our energy in this way, it is the same as if we received more of it, and this we can utilize to good account.

THE PHYSICAL FACTOR

Personal Magnetism depends upon various factors.

First of all sound physical health is essential. Without it there is little virility, and upon the presence of this vital stamina success largely depends. Theodore Roosevelt's dominating personality was due largely to his extraordinary physical energy. Large muscles are not necessarily a sign of this. It is the *vital* constitution which must be strengthened, and, in order to accomplish this, the internal organs must be in a healthy condition. Proper exercises, devoted to stimulating their function, should be taken for a few minutes daily; and in this connection the student would do well to consult one or two good books on physical culture,—giving directions of this character. Bending movement of all kinds are especially helpful. Deep-breathing exercises, which tend to expand the lungs, chest and diaphragm are to be recommended, and if you can stimulate the solar-plexus and internal organs by deep-breathing exercises, this will go a long way toward rousing the vital currents of the body. The inner psychical causes for this will be explained more fully in subsequent chapters.

THE MENTAL FACTOR

Next, the mind must be trained and cultivated in certain directions and channels. Just here the student would do well to turn back and re-read the directions given in Chapter VII, "Self and Soul Culture," where practical advice on success and its attainment is given. The practice of Concentration (Chapter XXIV) would prove very helpful here; relaxation both of body and mind should follow this.

The improvement of memory by various methods would greatly add to the strength of the psychic personality, since it is upon memory that the thread of personality depends. Attention upon any given subject should be cultivated, and you should never allow yourself to perform any action automatically which should be conscious. For instance, if you put an object in the drawer of your desk, make a conscious mental note of this at the time, so that you afterward remember where it is placed, and never allow yourself to place the object there without paying particular attention to it. Many people do this, and it is indicative of a weak power of attention and a scattered mind. The degree to which you can overcome this indicates concentration, and hence power. Nothing gives power and strength to the mind so much as continued exercise and concentration.

THE SPIRITUAL FACTOR

Spiritual development will also assist in the cultivation of personal magnetism, by drawing to your aid certain spiritual energies which *recharge* you:—that is, charge your body in much the same way that an electric motor is charged by external energy. This power you *draw* by placing yourself in a certain receptive condition which invites its influx.

All negative thoughts tend to erect a wall between yourself and helpful external guidance, and, on the other hand, an affirmative and positive attitude will have the effect of attracting or drawing to you this additional power.

Thoughts and emotions also have this effect. If you will carefully analyse your own inner sensations while thinking certain thoughts or experiencing certain emotions, you will find that selfish, self-centred impulses tend to contract you mentally and physically. You feel yourself tightening-up all over, as it were, and this internal action shuts off all outside aid and influence. On the other hand, if you think thoughts of friendship, love, etc., you will find your being tends to expand, and it is this feeling which opens the gates of your soul to an influx of higher power.

HOW TO INFLUENCE OTHERS

Personal Magnetism is practically useful in the affairs of this life. If you wish to achieve a certain object, you will be far more likely to do so if you have a good magnetic personality than otherwise. The following simple rules, if followed, will probably greatly assist you in the development of personal magnetism:

1. Just before entering into the presence of the person whom you are about to interview, call up that person's image before your mind, and assume toward it a positive mental attitude. If you do this you will carry over and maintain this attitude toward that person when you meet him. If you assume at the outset 60 or 75% of the mental dominance or initiative, you (figuratively speaking) only leave the other person 40 or 25% of the ground lying between you, which he can possibly occupy! Your business is to assume at the outset as large a percentage of the positive relationship as possible, and by doing so, you force the other person to assume the minor quantity.

THE USE OF THE EYES

2. When in the presence of the person whom you are to interview, look him squarely in the eyes, and hold his gaze and attention until you have won your first point. If possible, do not allow his attention or his eyes to wander from you until you have thoroughly insured his interest and sympathetic co-operation. It is important to catch the eye at the moment you are making a particular point, so as to "drive it home" as it were. You cannot stare a person in the eye all the while you are talking to him, and you should look away part of the time,—when you are discussing unimportant points or leading up to the climax. Many salesmen utilize this principle in making a sale. They will draw attention to a book or an illustration, at which they ask you to look, and talk about it for a moment; then close the book and make a short, quick remark, which will draw your attention to his face and eyes spontaneously. At that moment when he has gained your full attention,

and you are in a condition to receive any statement he will make to you, he will come to the climax of his argument and perhaps ask you to sign a certain paper, which you may be prevailed upon to do, under the influence of his personality.

HOW TO DEVELOP THE MAGNETIC GAZE

The eyes, therefore, play an important part in the cultivation of Personal Magnetism, and you should cultivate and strengthen them by certain exercises which will certainly develop them. For example: practise gazing steadily at an object for several seconds without allowing the gaze or the attention to wander, and without blinking the eyes. At first you will probably be able to do so for only a short time; but this will gradually be extended as you cultivate the power. Next, practise gazing at a fairly bright object, and continue this until you can look at it for several minutes at a time without becoming affected.

When you look into the eyes of another person, do not look blankly, but *will* at the same time, and throw the whole force of your personality into your gaze,—feeling that you will influence that person to do as you wish. Naturally, practices of this character can be, and, in fact, *are* utilized by many persons for evil as well as for good purposes. Those who are endeavouring to cultivate the higher side of their nature, however, will fully realize the necessity of utilizing any added powers they may gain for good purposes only.

PASSES AND SUGGESTIONS

3. Downward passes, as before explained, are sleep passes, and a few of these will add emphasis to your

speech and impress the person to whom you are talking. Do not gesticulate overmuch, however, as this will detract rather than add to what you have to say. A few passes at the proper moments will prove of great value.

4. Do not speak hurriedly, for if you do you will give the impression that you are in a hurry, and your hearer will unconsciously grow impatient. On the other hand, do not drawl your words, but speak naturally with a clear, forceful enunciation. The more reposeful and calm you appear, the more receptive your listener will be to hear what you have to say. At the same time, you must be business-like and precise.

HOW TO PREVENT THE INFLUENCE OF OTHERS

If you wish to offset the influence of some one, who is speaking to you, and prevent yourself from being influenced by him, you should see to it that you do not allow him to catch your eye at the psychological climax of the conversation, but studiously look away at that time, and carefully think over and analyse what he is saying to you, without allowing yourself to be swayed by his manner or words. Look at him in the intervals between these climaxes, when he will probably be looking away from you. Hold your mind in a positive attitude, and never allow yourself to be hurried into anything! The ability to say "No" and stick to it, when occasion demands, has been declared one of the greatest essentials to success, by many men who have attained great eminence. As Abraham Lincoln once remarked: "Be sure you are right and then go ahead!" A clear mind and inner mental repose will greatly add to your power in these directions.

HELPFUL APPLICATION

These exercises in the development of Personal Magnetism will be found especially helpful to all psychics, for the reason that they tend to offset and counterbalance, to a great extent, the subjective practices of mediumship, and hence balance-up the personality by accentuating the objective as well as the subjective side of one's inner self. All those who are developing psychic powers and mediumship should, therefore, while leading their daily lives, endeavour to follow the principles herein laid down, and develop their own natures along these lines. They will find that it will prove very helpful to them and preserve "that just balance we term health."

CHAPTER XXXII

PROPHECY *VS.* FORTUNE-TELLING

THE subject with which this chapter deals is a very important one for the Spiritualist, for the Psychic, and above all for the Public Medium, for the reason that it concerns him in a very practical manner.

It would seem as if Spiritualism, although an organized religious body, international in scope and influence, had no standing in the eyes of some people nor that its accredited mediums were entitled to any more consideration than ordinary "fortune-tellers." Fortune-Telling (so-called) is against the law, and in many cities the authorities are very severe on anything which can in any way be construed as fortune-telling.

Truly, one may be pardoned for believing that there is a power back of it which is opposed to so-called "Modernisms"—to the several movements of a spiritual and religious nature that are freshly putting forth real knowledge of our true relations to this life and the life beyond. It is not merely a moral wave, not merely ignorance of the difference between true and honest mediumship and fortune-telling, but an effort to retard and crush the truth. From the present standpoint of the court, Jesus, when he told the woman at the well about certain matters in her life, was a "fortune-teller." The people marvelled over Him because of what He could tell and do. To Spiritualists He was a *medium*, but a Master, and one so qualified by time and distance as

He comes down the centuries to the present age. In the 21st Chapter of 1st Corinthians, Paul describes the gifts of the Spirit (or spiritual gifts) and says they are all of the same spirit. The word *spirit* here is used in the sense of a collective noun or noun of multitude —much as we use the word Congress—and applies to the spirit world as the source of inspiration and control, the same as with the Spiritualist.

MEDIUMS AND THE LAW

There was much consulting with "mediums" in those early days of the primitive church; for, does not Paul again say, "Try the spirits and see if they be of God." "Prove all things; hold fast to that which is good." Opposition stirs up opposition and puts men and movements on the defensive. Spiritualism realizes this and is now actively engaged in efforts for the better protection of its mediums. When one strikes a blow at Modern Spiritualism, he strikes a blow as well at ancient spiritual truth—that Truth which fills the pages of our Bible, for which the early martyrs died and upon which the Christian Church was built. It comes as the Comforter which Jesus said he would send in the "latter days."

An assistant district attorney once made a ruling that a sandwich constitutes a meal, and so liquor could be bought on Sunday; but no Court can rule that a "fortune teller" constitutes a "spiritualistic medium" and have it stand; "The letter killeth, but the spirit maketh alive." At the same time Prophecy is a genuine spiritual or mediumistic gift, and there are thousands of persons who have experienced so-called premonitions or

previsions of the future, and have felt compelled to
tell others what they have seen for them.

Between "Prophecy" and "Fortune-Telling" there
is, therefore, a very fine line to be drawn, for the one
is dependent upon superstition to a great extent, while
the other is a genuine psychical faculty which requires
our recognition and study.

WHAT PROPHECY IS

So far as we can define the distinction between the
two, it may be said that prophecy depends upon internal spiritual promptings, or the reception of definite
messages relating to the future which are told the medium by external spiritual intelligences. He acts
merely as a medium for transmission in the latter case
and simply gives out what he receives. This is the type
of spiritual premonition, as distinct from clairvoyance
of the future, which we have already discussed in Chapter XIV. In this latter case the power appears to
depend upon internal and spontaneous quickening of
spiritual faculties and seems to be self-originated, as it
were. It is very similar to spontaneous premonitions,
therefore; and, in fact, these subjects are so very closely
connected that only an expert can define the differences
between them.

Unless one has had considerable experience and knowledge in this field, he is totally incapable of judging
whether a given set of phenomena are of the type of
genuine "prophecy" or mere "fortune telling," and he
should study the subject thoroughly before he is capable
of expressing an opinion upon it.

It may be well to consider the meaning of the word

"prophecy." It is derived from the Greek word, prophemi; *pro,*—meaning "before," and *phemi* "to say or tell." There is another word, *propheteuo,* of similar import and derivation, and means, to prophesy, divine, foretell, predict, presage; to explain or apply prophecies. —In Greek classical literature, the word prophet meant, a declarer, foreteller, diviner, a harbinger, a forerunner, a priest, teacher, instructor, interpreter; a poet, a bard. All of these definitions carry with them something of the idea of a character whose mission is in some way connected with the aspirations and longings of mankind.

A DEFINITION OF PROPHECY

The Standard Dictionary has defined prophecy as follows:—

1. To predict or foretell, especially under divine inspiration and guidance; to prefigure, as to prophesy evil.

2. To speak or utter for God.

3. To speak by divine influence, or as a medium of communication between God and Man. Specifically: To speak to men for God; declare or interpret the divine will.

4. To predict future events by supernatural influence, real or professed: To foretell the future; utter predictions, as, to prophesy a disaster.

5. (Archaic) To interpret scripture; explain religious subjects, preach; exhort.

Under the head of Synonyms, the Standard Dictionary gives: "Augur, divine, foretell, predict, prognosticate. Prophecy differs from predict by assuming a claim to supernatural or divine inspiration. To prognosticate is to predict from observed signs, indications, or conditions. To prophesy in the scriptural sense is to utter religious truths under divine inspiration, not simply always to foretell future events, but to warn, exhort, comfort, etc., by special message or impulse from God.''

This scriptural definition seems well adapted to the spiritualist sense of the word, when we interpret God to mean the Infinite Spirit of Good. The verb prophesy is also used in the New Testament in the sense of revealing something which had happened and was unknown to the person revealing it except through some so-called supernatural source; as for instance, after Jesus was pronounced guilty of death by the high priest, some of the ruffians, who have their counterpart in this day, spat in his face and buffeted him; and others smote him with the palms of their hands, saying:—''Prophesy unto us, thou Christ,—Who is he that smote thee?'' Matt. Chap 26 v. 65 to 68.

Jesus ignored this challenge. Could they have understood or would they have believed in his mission if he had correctly pointed out the man who had assaulted him?

EXPLANATION OF ''FORTUNE-TELLING''

It is true, however, that the method of arriving at the knowledge given is, in itself, an indication of the character of the knowledge imparted. Thus, fortune-telling, in the hands of charlatans and quacks, is often connected with such superstitious practices as reading the future from tea or coffee-grounds; from cards; allowing birds to pick out envelopes, containing written messages relating to the future, etc. Such practices are certainly to be deprecated by every sincere spiritualist and truth-seeker, though it should be said, just here, that many psychics who read the cards in this manner, depend not so much on the actual fall of the cards as upon the psychic impressions which they receive at the time the sitter's fortune is being told. This is often true, also, in the case of palmists. There is doubtless some truth in the general doctrine of palmistry, but it can only hold good to a very limited extent. When impressions are received, the process is somewhat akin to crystal-gazing, where the mind is concentrated on an external object while it remains passive and open to internal impressions; but, instead of receiving these in the form of visual pictures, they are given in a more general and vague manner.

WHY FORTUNE-TELLING IS SOMETIMES TRUE

On the other hand, genuine mediumistic messages are frequently given while the subject is reading the cards, examining the sitter's palm, etc. It will be observed that, in these cases, there is a certain fundamental reality in the phenomena, but it is perverted and unconsciously covered up by the seer who is unaware of the actual source of the information he gives. Psychic power or mediumship is the basis of the supernormal information given, but it is under the guise of fortune-telling.

A far more direct and satisfactory method would be to come out in a straight-forward and direct manner, and state that such and such impressions were received, relating to the future, and this premonitory faculty could doubtless be cultivated by certain practices and be used as the student progressed in his psychic development. Exercises for development of these faculties will be given later on, in this book.

Disbelievers in spiritualism often say: "If your assertions are true, why do not the spirits warn and advise you more frequently, and why do they not help you financially or otherwise, more than they do?" The answer is simply, as before said, that you are not a creator, but an instrument. A knife may be sharp, but it could not cut bread without a power behind it. A soldier may go to war and fight bravely without knowing the real reasons for the war. You are the knife or the soldier. You cannot act by yourself or achieve desirable results unless the power be imparted to you from beyond, and even then the power is supplied for other purposes and centred upon other things. The knife does not cut itself, but the bread. Clairvoyant power does not benefit the clairvoyant directly, but some third person, and, in cases where the student has found it possible to pervert its use and turn it into selfish channels, the power has invariably been lost. It may also be said that Spiritualists may err in their selection of spirit advisors as well as in their means of intercommunication. "That is true, for we are not endowed with perfect judgment even in selecting in this life our medical or legal advisors, or our governmental representatives and officials, our business partners or our friends, or the person to advise us as to where we can get the best advice in a given matter. The Spirtualist merely claims the right to act for himself without let or hindrance from those who differ with him in religious views. If he makes mistakes which cause him loss or suffering, it must be remembered that even Jesus, with his extraordinary psychic powers, made a mistake when he selected Judas Iscariot as one of the twelve. If it be said, that this seeming mistake was a part of a divine plan, then it may also be said, that the Spiritualist's seeming mistakes may also be a part of a divine plan."

HISTORY OF PROPHECY

There can be no doubt that prophecy has existed in all ages and has had its own uses as well as its abuses. Many spiritualists believe that prophecy is invariably connected with spirits and that the explanation depends upon their communication. On the other hand, many orthodox religious persons believe that prophecy depends entirely upon the influx of the divine spirit, and that the ability to predict or foretell comes directly from God. This is the manner in which it is regarded by many people and in many religious books. There are many references to prophecy and to prophets both in the Old and the New Testament, and any one who accepts the teachings of the Bible, as in any way true and valuable, can hardly fail to believe that prophecy is a genuine psychical faculty which has been exercised by men in all ages and is undoubtedly being exercised by them now. Thus: in 1st. Corinthians, Chap. 14, v. 3, we read: "But he that prophesieth speaketh unto men to edification, and exhortation and comfort." Again, in the same Chapter, v. 1, we read: "Follow after charity and desire spiritual gifts, but rather that ye may prophesy." And again in the same Chapter, verses, 31, 32, and 39, we read: "For ye may all prophesy one by one, that all may learn, and all may be comforted. And the spirits of the prophets are subject to the prophets. . . . Wherefore, brethren, covet to prophesy and *forbid not to speak with tongues."* . . . One more quotation: In 1st. Corinthians, Chap. 12, v. 4–12, we read: "Now there are diversities of gifts, but the same spirit, and there are differences of administrations, but the same Lord, and there are diversities of operations, but it is the same God which worketh all in all, but the manifestation of the spirit is given to every man to profit with all. For to one is given by the spirit the word of wisdom; to another the word of knowledge by the same spirit; to another faith by the same spirit; to another the gift of healing by the same spirit; to another the working of miracles; *to another prophecy;* to another discerning of spirits; to another diverse kinds of tongues; to another the interpretation of tongues; but all these worketh that one and the self-same Spirit, abiding in every man as he will." Many other references of this character could be given, but it is hardly necessary, for every student knows that every religious book in the world accepts the genuineness of prophecy and, in fact, all religions are based on the revelations of seers or prophets!

HOW IS PROPHECY POSSIBLE?

Prophecy is a faculty which usually comes unsought and spontaneously. When the future is seen in an isolated picture or event, it is usually called a premonition or pre-vision, and many examples of this character have been collected and published by the Societies for Psychical Research. It may be asked: "How is it possible to see into the future, to lift the veil of futurity and glance forward as we glance backward in reading history?" Certainly, at first sight such a thing appears not only impossible but absurd. Nevertheless it is an undoubted fact, and numbers of cases of this character might perhaps be explained more or less rationally—even with our present knowledge.

Thus: certain types of premonitions relate to the future welfare of the body or health of the subject experiencing them. In such cases we might suppose that the subconscious mind, which has a wider range of inner experience and knowledge than the ordinary waking mind, was aware of certain internal changes and happenings of which the conscious mind was totally ignorant. In such cases the explanation would be that this subconscious mind, having acquired this knowledge, would merely impart or "externalize" it in the form of a vision, voice or message, or in the form of automatic writing, etc.

A second type of premonition might depend upon subconscious inference and deduction,—this being far more acute and far-seeing than the conscious mind in such matters, particularly when the latter is occupied with every-day practical affairs.

Another set of premonitions might be accounted for by assuming that the knowledge given is imparted telepathically or gained clairvoyantly by the subject's own mind. In these cases the information would be in the minds of other living persons and would be gained from them and given out before the subject had gained the fact normally.

SCIENTIFIC EXPLANATION OF PROPHECY

A fourth type of premonition might be explained by assuming that discarnate spirits play a large part and communicate the information to the recipient of the message in question. In this case the discarnate intelligence would have to be in possession of certain facts or be enabled to see further than the psychic himself, and there is much evidence that this is in fact the case, on numerous occasions. For example, if we see a spider walking across the table, we know that when it reaches the edge, it will either stop or fall over, though the spider cannot foresee these facts and continues to walk quite ignorant of the fate in store for it! Again, to use a more forceful example, supposing a friend of yours is walking down the street and is coming to a cross-street down which a strong wind is blowing. Being in possession of this knowledge, you can predict with more or less certainty that when your friend reaches this cross-street, his hat will be blown off, and, in fact, this actually happens. Now, you will see, in this case your ability to predict this fact (or partly see into the future) was based on your *larger knowledge* of certain factors playing about his life. It is only logical to suppose, therefore, that spirits who may be and probably are in possession of greater psychic powers than we, can foresee tendencies and destinies, to a certain extent, towards which human beings are tending. This being so, they are enabled at times to communicate (perhaps telepathically) statements regarding the future which often turn out to be true. This would be a logical explanation of many cases of premonition of this type, and would explain to us, in a perfectly simple manner, why it is that mistakes and errors so often occur in premonitions of this kind. It would be only what we should expect.

It must be admitted, however, that there are many cases of premonitions which cannot be explained in this simple way and which we cannot in any manner account for, in the present state of science and of our limited knowledge of psychic phenomena. These cases we must simply record and hope that the time may come some day, when we will be enabled to comprehend clearly the underlying causal explanation which will make clear to us the real mechanism by means of which premonitions and prophecies are fulfiled.

CHAPTER XXXIII

REINCARNATION AND EASTERN PHILOSOPHY

Most religious philosophies of the East are based on the reality of reincarnation, or the embodiment of the same soul in a variety of physical bodies, living on this earth at various stages of the world's history, often separated from each other by a number of years. The doctrine contends that the same individuality is maintained throughout all these lives, as a "background," but that each life is also an individual experience which is destined to teach the soul one or more particular lessons, which it needed to learn for the purposes of its ultimate progression and perfection. The doctrine is based largely upon the Law of Compensation, which says that, inasmuch as there is, in this life, so much obvious inequality as regards the material returns, rewards and happiness, there must be another chance for that soul at another time and under other circumstances, and that the poverty and other conditions which may be present in this life are for a purpose, and teach a lesson, and that quite possibly in some past life, the same individual has been extraordinarily wealthy and has misused the riches and power entrusted to him.

THE MEMORY OF PAST LIVES

This is a fascinating doctrine and one which, at first sight, asks us to yield our consent to it; yet there are many objections to this theory of reincarnation, as we

shall see presently. It may be well to answer here one
main-objection to the doctrine which is sure to be advanced by the ordinary critic, and this is, that, if the
same soul be reincarnated a number of times, it should
remember its past lives,—while as a matter of fact it
rarely does so, and if we are to profit or benefit by these
in any way, one would think that this memory would
be absolutely essential.

The Theosophist or Reincarnationist replies to this,
however, by stating that each life is intended to be an
individual, separate existence, without a memory-bond,
or connection with any previous life. The soul of the
individual which reincarnates only reaps the knowledge
of each life *after death*, when this knowledge is added
to the total mass of experiences already gained. Thus
the individual human life is conceived to be greater
than any single life, just as a bucket of water is composed of thousands of drops, each drop being separate
until merged with others into the whole. In the same
way each individual life, representing a separate drop,
would be individualized until after death, when it is
again merged into the total personality.

Our inability to remember former lives is accounted
for by assuming that there is no direct connection between the *total* self and the self which is built up in
this life through the physical brain. They are separated, though it would take too long to explain here
exactly the nature and causes of this separation, according to the doctrines advanced.

THE ARGUMENTS FOR REINCARNATION

Another argument, which is advanced in favour of
the doctrine of reincarnation, and to many minds a very

strong one, is that life must necessarily be eternal and immortal, inasmuch as it is indestructible by death, and continues to exist to all eternity in the future, and, for the same reason, it must also have existed from eternity in the past, and it is "inconceivable" that such a thing as an individual human spirit should continue to exist for ever *after* the moment of birth, while it did not exist at all *previous* to that event.

These are the main arguments, which are brought forward in favour of the doctrine of reincarnation, and we may add to these one argument based upon experimental evidence. It is this: that many of those who have progressed sufficiently in their psychic development can—so they say—remember their past lives,— either fractions of them or incidents in them, or the whole life may be remembered as a consecutive series of scenes and events. Many of the leaders of Theosophy and other religious systems of this character contend that they can actually do so. The majority of spiritualists, however, are opposed to this view, and contend that reincarnation is not a fact, though it must be admitted that in the past there has been a great diversity of opinion on this subject. The French School of Spiritists, formerly headed by Allan Kardec, contends that reincarnation is a fact, and Kardec's work "The Spirit's Book" is based entirely upon teachings of spirits, who claim that reincarnation is true! On the other hand, the majority of German, English and American mediums contend that reincarnation is not true, and spirits who return through them also assert emphatically that it is not a fact! The reason of this apparent contradiction was explained in an earlier chapter. (The communicators merely stated their own views and opinions.)

Now, in considering this doctrine of reincarnation, there are certain factors which we must bear in mind:

1. The average scientific inquirer begins by doubting the reality of survival at all, and contends that nothing persists after the change called death. For him it is annihilation! The first point to be proved, therefore, is that anything at all exists after death, and the phenomena of spiritualism are the only ones which prove this, as before pointed out. Until it is thoroughly established that "spirit" of any character continues to exist after death, it is useless to argue whether or not such a spirit is "reincarnated," for the reason that the average sceptic would contend that there is no such thing as a spirit to reincarnate! Until this primary fact of spirit existence is proved, therefore, it is useless to argue concerning this question of reincarnation.

2. Assuming that this is granted, still there is no proof that reincarnation is a fact, if we demand "proof" in the scientific sense of the world. In order to establish such a doctrine as this, a tremendous mass of testimony would be necessary,—far more than the ordinary phenomena of spiritualism, which claim to establish a comparatively simple truth. Yet, as a matter of fact, there is far *less* evidence, as we all know, for the reality of reincarnation than there is for spirit-return. As "the strength of the evidence should be proportioned to the strangeness of the facts" it will be seen that we are as yet very far from proving reincarnation according to this standard. A vast mass of well-attested evidence would have to be forthcoming, and this has not been produced.

LIFE, PAST AND FUTURE

3. It is not necessarily true that because the human spirit continues to exist for all eternity in the future, it must necessarily have existed from all eternity in the past. Physics teaches us that a body set in motion comes to rest because of the hindrance or friction from outside forces acting upon it. If there be no friction to retard such a body it would, theoretically, go on for ever in a straight line. Once give a ball an initial push, and, provided there is no friction, it would roll on for ever without coming to a stop. It might well be, therefore, that the human spirit once initiated, would continue in the same fashion, since we can see no hindrances to its progress resembling those acting in our physical world. Again, a speck of mud thrown off from a revolving wheel only exists *as* an individual speck, after it was thrown off in this manner; before, it was a part of the general mass.

Assuming, therefore, that an individual human spirit is in some way separated and individualized at birth from the general stock of cosmic life-energy, at the moment of conception, it might be that it continued as an individual thing thereafter for all eternity, without necessarily having existed *as such* in the past.

THE SPIRAL OR VORTEX OF LIFE

In the next place, assuming that life is an individualized force, we can quite conceive that this force, ascending in a series of spirals, tends to become more detached and individualized with each revolution through which it passes, and that ultimately it will tend to become detached and thrown off, as it were, from the

vortex of life as an individual being.   Birth might represent this process; and again we see that it is not necessary to suppose that human spirit must have existed in
the past because it continues to exist in the future.

As to the law of compensation, already mentioned,
this is not really an argument but rather an emotional
belief, based upon the idea of justice.   But, in the first
place this may not necessarily be true; and in the second,
even supposing that it is, the same result is reached in
other religions, for according to the teachings of orthodox Christianity the reward of the poor but righteous
is in Heaven, and according to spiritualistic philosophy
it depends on individual progress and effort.

HOW WE REMEMBER PAST LIVES

The doctrine of reincarnation cannot, therefore, be
said to present a logical justification for the belief.
There remains the more substantial foundation, based
upon the before-mentioned experimental proof, namely,
that many persons claim that they can remember portions of their past lives and even that they can remember the whole of them.   These latter cases, however, are
very rare, and the material from which one could form
one's judgment regarding such cases has never been
published.   Owing to the lack of respectable evidence in
this direction, therefore, we may assume for the present,
and until proof to the contrary be forthcoming, that
such cases depend not upon reality, but upon elaborate
subconscious imaginations and romances which these
individuals have constructed within themselves, as the
result of brooding and thinking over possible past lives
of their own.   There are many analogies for this belief,

and in some cases at least, it has been proved beyond all question of doubt, that these "past lives" were in reality fictitious, and that the "memory" of them, so-called, was certainly and purely subconscious imagination. Those who may be interested in obtaining this proof are referred to Professor Flournoy's book "From India to the Planet Mars."

WHERE AND HOW THESE "MEMORIES" ORIGINATE

There remain those cases, far less satisfactory and convincing but far more numerous, in which isolated incidents of past lives have been remembered, or in which scenes have flashed up before the mind, together with the impression, amounting to a certainty, that the individual has experienced or lived through that scene before. Most cases of this character may be explained in a perfectly natural manner, and do not afford any direct proof of the doctrine of reincarnation.

Let me explain a few of the causes which may be operating, inducing such apparent "memories of past lives."

In the first place, many of them are due to so-called illusions or hallucinations of memory—so-called "pseudo presentiments," in which the event and the feeling that it has transpired become reversed or transposed in the mind, so that one remembers the impression as occurring *before* the real event, while in reality it happened *afterward*. That this occurs in many cases has been scientifically proved.

In the second place, dreams or subconsciously noted impressions which never come into consciousness, may suddenly flash up, in connection with a certain mental event, and this would give rise to a feeling (true, in a sense) that we had experienced it before. We had, but in a dream, and not in a previous life!

Thirdly, many experiences, conversations, etc., overseen or overheard before the age of four, when the personality is in the process of formation, and when consecutive memory and consciousness of "self" is said to begin, may be remembered as isolated experiences, and these may also give rise to the impression that we had seen them, or experienced them before. Again, this is a fact, but it was not in a "previous life."

Lastly, there are many cases in which the subconscious mind noted a scene or event a fraction of second, or perhaps several seconds, or even minutes, before the conscious mind did; and when the latter became aware of it, there would again be this sense of "familiarity," and the feeling that we had seen or experienced this event before. This is true, but it was only a short time before the actual experience.

For all these reasons, therefore, and others which it would take too long to give,—the majority of spiritualists and "psychical researchers" do not at present regard the doctrine of reincarnation as true, or in any way adequately proved, and prefer to believe, until this proof be forthcoming, that the individual human spirit is initiated at birth, builds up its own life by its own efforts and experiences, and that it continues to improve upon this life, by continuous striving, after it has reached the spiritual world, in the same manner that it does here on earth.

CHAPTER XXXIV

THE ETHICS OF SPIRITUALISM

As explained at the beginning of this Book of Instruction, Spiritualism is not only a scientific question, but it is also a philosophical and a religious question. It is approachable from the point of view of phenomena; also that of theology and ethics. The student who has followed the work thus far has doubtless progressed to some extent in the understanding, if not in the control, of psychic phenomena, and fields of knowledge have been opened up before him, of which he had previously been more or less unaware. But all this would not only be unavailing, but harmful, if Spiritualism were not ethically and spiritually right as well as phenomenally true. It is no good developing something which leads one ultimately only into a mire of harmful results and a false philosophy. If Spiritualism cannot be justified from the religious and ethical standpoint, it should be let entirely alone by all save the few scrupulously scientific investigators who approach the subject from that point of view and not as a belief. It is very important, therefore, for the Spiritualist to have his belief founded in correct ethical principles, for, as I have before pointed out, the reproach has been raised against Spiritualists that "they are everything but spiritual." Unfortunately there are many of this type, but they are doubtless in the minority, and the majority of Spiritualists wish to see their faith grounded on firm ethical principles.

IS IT RIGHT TO INVESTIGATE PSYCHIC PHENOMENA?

Various questions arise in this connection:

The objection to Spiritualism may first of all be raised that "such things are God's secrets which He keeps to Himself. What is the use of seeking? You will find nothing!" But to this Monsieur Camille Flammarion replies rightly: "There always have been people who liked ignorance better than knowledge. By this kind of reasoning (had man acted upon it) nothing would ever have been known of this world. . . . It is the mode of reasoning adopted by those who do not care to think for themselves and who confide to directors (so-called) the charge of controlling their consciences." If these phenomena really exist they must be part of the universe and subject to natural law, for otherwise they could not exist at all. There is no such thing as the "super-natural." All is natural, even if it be the communication of spirits. It may be unusual or uncommon, and because of this we call these phenomena super-normal, that is beyond the ordinary normal experience of mankind; but they are not and cannot be super-natural.

CONCERNING FRAUD AND ERROR

Again the objection may be raised that these phenomena "foster superstition," but this is based upon the belief that the phenomena are necessarily *untrue*. Once the reality of the facts is established, there is no "superstition" connected with it. It becomes merely a question of scientific evidence.

Again the objection has been raised against spiritualism on the ground that it encourages fraud and charlatanism. To some extent this is true, but other cults have suffered in the same way, and all sincere spiritualists are the first to expose falsity and fraud when they meet it. There are Spiritualists, it is true, who endeavour to shelter fraudulent mediums and pretend that this fraud does not exist. Such a method is a great mistake and only tends to degrade and lower Spiritualism as a religious belief in the eyes of the public. "Truth is mighty and shall prevail," and Truth should above all else be the watch-word of the true Spiritualist.

IS IT HEALTHY AND NORMAL?

Then there is the objection that spiritualistic practices encourage morbid and abnormal states and conditions and help to induce insanity. Again there is some excuse for this argument; but as so often pointed out, it is the conscious or unconscious *abuse* of psychic and mediumistic power rather than its *use*, which is so dangerous and detrimental. In the initial experimental stages of Spiritualism, some harm has doubtless resulted to some experimenters, but this is only a stronger reason for urging us to discover and rightly understand the laws and conditions under which psychic phenomena and spirit-communication may operate. When these are once understood, they are thereby rendered safe, and thenceforward there is no reason why spiritualistic practices should be unsafe—save for those who neglect its well-ascertained laws.

Again it has been urged that it is wrong to communicate with spirits of the departed for the reason that such communication is "not natural" and that by doing so we interfere with the progression and spiritual development of those who have passed over. But the reply to this is two-fold: In the first place, the many cases of spirit-return which are recorded prove that these phenomena are far more common than is usually supposed, and for this reason it is not so exceptional a thing, but almost a common occurrence. It partakes more of the nature of natural law than of an experimental or miraculous event. If such is the case, it can hardly be detrimental or unnatural, since none of nature's laws are unnatural.

DOES SPIRIT-COMMUNICATION RETARD PROGRESSION?

Again, there is no reason to suppose that communication retards the spiritual progress of those who have died—on the contrary, we might suppose that, in many cases at least, such communication would certainly *help* the spirits; and in many cases, as we know, they have repeatedly come back for the express purpose of asking the living to carry out some mission for them which weighed upon their minds, and they have stated that they could get no rest or comfort until this mission has been fulfilled. There are many cases, again, as we know, wherein the returning spirits have requested help and the prayers of the living to assist them in their progress, and many Spiritualists have devoted their lives to this work—namely, assisting earthbound spirits and helping them in their natural spiritual progress. Many spirits have returned to impart certain information, or to give counsel, warning or advice to friends and relatives of theirs, still living; and we cannot but believe that they are far happier in doing so than if they were

obliged to stand by and see some unhappiness, accident
or catastrophe overtake their loved ones on earth, while
they themselves were obliged to remain inactive. Were
they still alive they would like to feel that they had
prevented such a catastrophe, and it is only natural to
suppose that they continue to live in this way and
continue to take an interest in their loved ones after
they have passed over. In this way spiritual communication becomes a natural and beautiful belief.

SHOULD THE "DEAD" KNOW OUR SORROWS?

This brings us to another important question from
the ethical point of view, and this is that the so-called
Dead are in constant sympathetic communication with
those still living, and that they, after they have died,
have a knowledge of our lives, our trials and our tribulations. Many religious persons contend that this is
a very unethical belief and that they should know nothing of those on this earth after once they have died.
Yet, this is surely contrary to all human sympathy and
experience. A mother, wrapped up in the interests of
her child, would surely prefer to remain near it and
watch over, guard and guide it, if possible, for a few
years rather than to desert it wholly and be totally ignorant of its life and progress. Yet, this is what orthodox religion contends they should do! Spiritualism is
far more ethical in this respect than the ordinary religious teachings, since it tells us that constant, sympathetic *rapport* exists between this world and the next,
and that there is no abrupt severing of the ties of human
sympathy and love at the moment of death. This,
surely, is a comforting thought for the bereaved.

THE ETHICAL TEACHINGS OF SPIRITUALISM

The religious teachings of Spiritualism are otherwise far more ethical than those of any other religion. Instead of a world devoted to selfish personal progression, subject to the changeable whims of an external Deity, we have in the teachings of Spiritualism a perfectly consistent and scientifically founded religious faith, quite in accordance with the doctrine of evolution. All progress depends upon personal development. As Dr. Alfred Russel Wallace says in his "Miracles of Modern Spiritualism"—"The hypothesis of Spiritualism not only accounts for all the facts (and is the only one that does so) but it is further remarkable as being associated with a theory of a future state of existence which is the only one yet given to the world that can at all commend itself to the modern philosophical mind. . . . The main doctrines of this religion are: that after death, man's spirit survives in an ethereal body, gifted with new powers, but mentally and morally the same individual as when clothed in flesh. That he commences from that moment a course of apparently endless progression which is rapid just in proportion as his mental and moral faculties were active while on earth. That his comparative happiness or misery will depend entirely on himself, and that just in proportion as his higher human faculties have taken part in all his pleasures here, will he find himself contented and happy in a state of existence in which they will have the same exercise; while he who has depended more on the body than on the mind for his pleasures will, when that body is no more, feel a grievous want, and must slowly and painfully develop his intellectual and moral nature till its exer-

cise shall become easy and pleasurable. Neither pun-
ishments nor rewards are meted out by an external
power, but each one's condition is the natural and in-
evitable sequence of his condition here. He starts again
from the level of moral and intellectual development
to which he has raised himself while on earth.''

SHOULD MEDIUMS ACCEPT MONEY?

One other point remains to be considered. It is this,
that mediums accept money for their services, and inasmuch as this is a spiritual gift, it is wrong! Yet, this
is common to all other religions. Do not the ministers
of all other religions receive compensation in some
form or other for their services? As long as mediums
are living in this material world, they are obliged to
meet the costs of living like all other human beings,
no matter how spiritual their work or they themselves
may be. If mediums possess genuine power, it is only
natural, in a sense, that they should utilize it and turn
it to account, and it is certainly true that by doing so
they help their fellow men and help those who come to
them as much or more than men in any other walk in life.
This being so, it can hardly be said that any aspect of
Spiritualism is in itself unethical. It is, on the contrary, the most sensible, rational and ethical religion in
the world!

CHAPTER XXXV

WHAT HAPPENS AFTER DEATH?

PRECISELY what happens at the moment of death is one of the most dramatically interesting and one of the most striking, insoluble problems in the world today; it remains for us the problem of all problems, the mystery of the universe, the science of being. One moment we see a figure before us—a muscular, powerful man, capable of heroic efforts, great intellectual flights, lofty aspirations, delivering an oration which stirs the hearts of thousands and perhaps helps to sway the destiny of nations and change the map of the world! The next moment he is lying on the floor, a corpse, lifeless, inanimate, incapable of the slightest thought, the slightest muscular exertion. He is the victim of "heart-failure."

THE MYSTERY OF DEATH

A second and all has changed! Nothing can now be influenced by him; nothing is now possible but the gradual decomposition of the body and its return to the dust from whence it sprang. Could any change be more profound or more lasting—since it can occur, presumably but once in all eternity? The slightest anatomical variation in the man's body—so small, perhaps, that even a microscope cannot detect it—and we then behold the most mighty change which occurs in nature, the profoundest of tragedies and wonders! We behold the transition from the living to the lifeless, we pass from life to death. What is this change we have seen before

299

us? Can we in any way understand it? What can we learn? What see? These are questions over which men have pondered for centuries and which still form the most fascinating problem in the realm of spiritual inquiry.

WHY WE SHOULD NOT FEAR DEATH

Many persons fear death; but they should not do so for the reason that, on any theory, it is not a thing to be feared. It has been proved abundantly that, with the exception of very rare cases, there is no pain at the moment of death and no consciousness of dying. Both are obliterated by the kind hand of nature. The suffering which goes before belongs rather to life than to death, and, in fact, many of those who have suffered from some torturing disease have died with a smile of happiness and contentment on their faces. Of the *physical* body we need think little, since the spiritual body separates itself from this body after death, and thenceforward is as unconscious of it as we are of a finger or any part of our body which has been cut off during life. The human spirit takes some time to become severed completely from the physical body, and for this reason it should not be buried or cremated too soon after death. But with these exceptions, we need think little of the state of the physical body, since we sever our connections with it entirely as soon as we pass into the spirit world. What happens after we have effected this separation is naturally a question of absorbing interest to many minds, since all of us have to look forward to this experience.

The statements of clairvoyants and of those "spirits" who have returned to tell us of their passage into the next life should, therefore, be of considerable interest in this connection. Let us see what they have to say.

A CLAIRVOYANT DESCRIPTION OF DEATH

Andrew Jackson Davis, one of the founders of Modern Spiritualism, and a gifted seer, describes the process as follows:—"Suppose the person is now dying, it is to be a rapid death. The feet first grow cold. The clairvoyant sees right over the head what may be called a magnetic halo—an etheric emanation, in appearance golden and throbbing as though conscious. The body is now cold up to the knees and elbows, and the emanation has ascended higher in the air; the legs are cold to the hips and the arms to the shoulders, and the emanation, though it has not risen higher in the room, is more expanded. The death-coldness steals over the breast and around on one side, and the emanation has attained a higher position nearer the ceiling. The person has ceased to breathe, the pulse is feeble, and the emanation is elongated and fashioned in the outline of the human form,—beneath it is connected with the brain. The head of the person is internally throbbing—a slow deep throb—not painful, but like the beat of the sea. Hence, the thinking faculties are rational while nearly every part of the person is dead. Owing to the brain's momentum, I have seen a dying person even at the last feeble pulse-beat rise impulsively in bed to converse with a friend; but the next instant he was gone—his brain being the last to yield up the life principle.

FORMATION OF THE SPIRITUAL BODY

"The golden emanation which extends up midway to the ceiling is connected with the brain by a very fine life-thread. Now the body of the emanation ascends. Then appears something white and shining, like a human head; next, in a very few moments a faint outline of the face divine, then the fair neck and shoulders, then in rapid succession come all parts of the new body down to the feet—a bright shining image, a little smaller than its physical body, but a perfect prototype or reproduction in all, except its disfigurements. The fine life thread continues attached to the old brain. The next thing is the withdrawal of the electric principle. When this thread snaps, the spiritual body is free and prepared to accompany its guardians to the Summerland. Yes, there is a spiritual body—it is sown in dishonour and raised in brightness.''

HOW IT FEELS TO ''PASS OVER''

Hear, again, what a returning "spirit" says, who has passed through the "Valley of the Shadow of Death" and has apparently returned to tell us his experiences:— ''When I awoke in the spirit-life and perceived that I had hands and feet and all that belongs to the human body, I cannot express to you in the form of words the feelings which at that moment seemed to take possession of my soul. I realized that I had this body—a spiritual body. . . . Imagine, if you can, what the surprise of a spirit must be, to find, after the struggle of death, that he is a new-born spirit, free from the decaying tabernacle of flesh, that he leaves behind him. I gazed on weeping friends with a saddened heart, mingled with joy—knowing, as I did, that I could be with them and behold them daily, though unseen and unknown, and as I gazed upon the lifeless tenement of clay and could behold the beauty of its mechanism, I felt impelled to seek the Author of so much beauty and youth and prostrate myself at his feet. I felt a light touch on my shoulder, and, joy unspeakable! I beheld the loved ones of earth, some of whom had long since departed from the earth plane . . . and I felt myself ascending or rather floating onward and upward through the radiance of space. I saw about me many spirits and their guides bearing them company through the bright realms of immensity.''

NOVEL EXPERIENCES ON ''THE OTHER SIDE''

So the human spirit, issuing from the body, gradually rises higher and higher, and comes into touch and harmony with those about it, and with those who possess sympathy and mutual interest. As explained in the chapter devoted to the ''Spirit World,'' it is highly improbable that there are any *physical* barriers between the ''spheres,'' one from another; but they are doubtless separated, nevertheless, by walls of mental and spiritual origin. If we are in one of these planes, we must progress upward, before we can reach or remain with those whom we desire most; and for this reason there is a ''hell,'' so to say, for those who cannot attain what they desire,—which can only be by continual striving upward and onward. In this, however, they are constantly helped and assisted by spirit guides and helpers; so that progress is rapid, when it is really desired and worked for.

HOW WE TURN BACK TO COMMUNICATE

On the lower planes of existence communication with those of the earth is, it is said, comparatively, easy; but this becomes increasingly difficult as we ascend in the upward scale. It has often been pointed out that descending to communicate with those on earth is something like going down to the bottom of a muddy pool; and those who desire to go to the bottom of muddy pools are very rare, even on this earth! Still, spirits moved by ties of love for those left behind, make the attempt from time to time;—successful or unsuccessful in proportion to favourable or unfavourable conditions. This however we discussed in former chapters.

As we progress, we are said to acquire more interest in the new world; and lose interest in this—just as we gradually lose interest in one country when we move into another. New scenes, new interests, and the new environment gradually alter our line of thought; but just as we are always glad to see a relative or an old friend from the home town, or the "motherland," so are we most happy to meet and greet those who "pass over," when their turn comes to join us in the spirit world.

HOW WE PROGRESS IN THE SPIRIT WORLD

After these initial stages have passed, upward progress and development begin. For many, however, the shock of death has a very severe effect, especially in cases of suicide and those who have met with sudden and violent deaths. In such cases these spirits require some time to recover their normal selves, and have to be nursed back to health, as it were, on the other side. The same is true of those spirits who have had their minds affected by some mental or physical disease. But, after this stage has been passed, they all emerge into the brightness beyond, and begin their interest, their instruction, their learning and their progress of soul and spirit, as well as of intellect,—which is to occupy them for ages of time to come. Those who "die" are received and cared for by loving friends on the other side, just as they were when they were born into this world. One need have no hesitation nor fear on this account. Physical birth is a terrible experience, but we remember nothing of it; and there are always those present who will tender and care for us.

In the same way, birth into the spirit world through the gates of death is, in many ways, a terrible shock; yet we are cared for by loving guardians and received with love and care,—finding happiness awaiting us when we pass from this world into the world beyond.

CHAPTER XXXVI

BAD AND PERVERTED USES OF SPIRITUALISM

EVERY gift or power can be abused; many in the past have turned their increased psychic powers into evil channels (at various times in the world's history) and who continue to do so today. They are known as magicians, witchès, vampires, possessors of the evil eye, etc., etc. For the moment it may be pointed out that psychic unfoldment and increase of psychic power brings with it added responsibility. As our power in this direction is increased, so also we are expected to use it rightly. If much has been given us, much is expected! It is quite possible, it is true, for these powers to be turned to bad account, and others injured, wealth acquired, etc., temporarily by their use. But if these powers are used for these purposes, they are usually soon lost, and then the student is in a far worse condition than before, for the reason that he is not only without the added power which he craves, but also has deteriorated mentally, morally and physically as the result of their harmful use.

THE DIFFERENCE BETWEEN MAGIC AND MEDIUMSHIP

In the middle ages psychic powers were undoubtedly used for good and bad purposes. White magic was beneficial and black magic harmful. White Magic invoked "angels"; Black Magic invoked "devils." In neither case were the spirits of departed human beings called upon,—but rather intelligences, either lower or higher than man in the human scale of evolution.

Another thing which distinguishes mediumship from magic, is that mediumship is more in the nature of a request,—a calling upon human intelligences for help and advice. Magic, on the other hand, depends upon invocation or *demanding* the presence and assistance of other intelligences differing from the human, and their assistance in the work to be performed.

HOW INVOCATIONS ARE PERFORMED

For the purposes of this invocation, various magical practices were undertaken, such as prayer, the saying of certain words and sentences, preparation of the magic circle, with its pentagram, seal of Solomon, etc., as well as utilizing various magical preparations secured from dead bodies and the poisons of animals and reptiles, etc. These magical practices were usually undertaken at certain seasons and phases of the moon, after long training on the part of the magician, and in specially prepared rooms or localities, which had been kept apart only for magical purposes. Exact descriptions of such invocations and the methods employed are to be found in certain rare books on the Ritual of Magic; but inasmuch as they are neither healthy nor desirable, we do not deem it wise or right to place these teachings before the student, who might be tempted, did he possess the knowledge, to put them into operation, and thus injure himself mentally and morally—perhaps beyond repair. (Students who are interested may consult:—A. E. Waite, "The Book of Black Magic and of Pacts"; Levi, "The Doctrine and Ritual of Transcendental Magic," etc.)

AN EXPLANATION OF "WITCHCRAFT"

During the middle-ages, also, witchcraft flourished. It depended upon the use of certain psychic powers which witches were said to possess,—only in their cases this power came directly from the "devil" himself,—being bestowed upon them in person by his Satanic Majesty! The witches were all said to meet two or three times a year on some lonely mountain top at midnight, —these meetings being called "Sabbaths." At these sabbaths all sorts of magical and anti-religious ceremonies were held. The sacrament was mocked, the devil was worshipped, etc. The witch was said to swear allegiance to the devil, who thereupon touched her on some part of the body which became anæsthetic,— lacking all sensation. These marks occurred in various parts of the body, and such marks were consequently known as "witch marks." The probable explanation of such cases is that in connection with the abnormal mental and physical states induced by witches, there resulted a peculiar form of hysteria in which small zones or patches on the body became anæsthetic. Modern science now recognizes the existence of such insensible patches and calls them "anæsthetic zones." They are typical of this form of hysteria. This is the modern scientific explanation of the so-called witch-marks.

The journeys to the sabbaths were doubtless, for the most part, imaginary flights,—resulting from the administration of opiates and other drugs which they were known to take, and with which they anointed their bodies. At the same time it is probable that there were many genuine supernormal psychical phenomena connected with witchcraft, and this is becoming more and more probable as we progress in the understanding of such cases.

DEVIL WORSHIP

Another form of perverted occultism is that of "Devil-Worship," which exists in various forms even today in Paris, the Malay Peninsula, in London, in New York and doubtless in other large cities. At these meetings, which are devoted to Devil-Worship, various invocations, etc., are gone through, and the "devil" is said to appear in person and bestow power upon certain privileged members of the club who are thereafter enabled to use certain powers to their own advantage. Many of the scenes of these devil-worshipping societies are too revolting to be described; but have been pictured at length on one or two occasions by those who have taken part in these invocations.

THE "EVIL EYE"

Again, certain individuals have a power which is known as the "Evil Eye." This is particularly believed in by the peasants of Naples and Southern Italy, by the peasantry of Southern Spain, Austria, and other countries. Any one possessing the evil eye is supposed to have the power of bewitching or maiming any person or animal upon whom he throws his glance. Cattle, looked at by one possessing the evil eye, invariably become sick and die; crops fail, pestilence falls, etc. The evil eye is a gift which is usually unsought, but comes spontaneously and is not desired by any one. The sure way to guard against the evil eye,—according to the beliefs of the countries mentioned,—is to extend the first and fourth fingers of the hand toward the pos-

sessors of the evil eye—the second and third fingers
being folded over into the palm of the hand and kept
there by the thumb. In this position the outer fingers
somewhat resemble the horns of a bull, and if the hand
holding the fingers in this position be pointed at any
of the children or beggars in the above-named countries,
they will usually turn and fly from the sign-maker!
Many Europeans use this knowledge to rid themselves
of pestilent beggars!

VAMPIRES AND HOW THEY ATTACK

Another form of evil influence which is said to exist
and is particularly believed in by the natives of Silesia,
Moravia and Southern Carpathia is that covered by
the general word "Vampire." In our ordinary language a vampire is a species of bat, and the word is
employed because human vampires were said to assume
the shape of large bats at times, flying in the window
when their victims are asleep. A vampire is one who
sucks the life-blood of his victim, through two small
holes punctured in the skin, in very much the same
way that a mosquito sucks our blood after puncturing
the epidermis. These holes are said to occur usually
in the throat, and the victim is, of course, attacked
as a rule during sleep.

Those who are vampires after they are dead and
buried are enabled in some miraculous way, it is said,
to leave their coffins and tombs and wander about seeking victims. When they are dug up, they are found
fresh with a pink complexion and the whole body en-
gorged with blood. The only sure way to kill vampires,
it is said, is to drive a stake through the heart or cut
off the head, when a quantity of fresh blood will gush

forth and the vampire is killed for ever! Tradition also says that those who are bitten by vampires become vampires in turn.

MODERN VAMPIRAGE

Vampires of a certain sort, however, are not unknown in our own day. In an interesting article on "Vampires," in the *Occult Review*, June, 1908, Dr. Franz Hartmann described a method of what might be termed natural vampirage. He refers to the Bible (1 Kings, 1) and also alludes to certain processes by which one person is enabled to draw vital energy from another,—by establishing close contact. This process of Nature is governed by well fixed laws. Through ignorance of these laws, many people have become victims of "Modern Vampirage." Another form of perverted occultism which remains is the employment of charms, amulets, talismans, etc., which are often sold for the purpose of inducing mental and physical disease and "Black Magic" which has existed through all ages. We must not forget, also, the so-called "Voodoo" practices of the natives of West Africa, which are said to be remarkable by those who have witnessed them.

HOW TO PROTECT YOURSELF FROM OCCULT AND EVIL INFLUENCES

It is often a little difficult for the modern student of the occult to determine just how much he is to believe in these stories. Undoubtedly most of them are based on superstition, fanaticism and imagination. At the same time there is enough truth in them to make us be cautious and put us on our guard. Never, under any circumstances, should you undertake to practise any

of them for low, selfish purposes. In order to protect
yourself from influences of this sort, if you feel that
they are being wielded against you, resort to the measures outlined in previous chapters and you may be sure
that if you do this, you will be impervious to all ordinary influences of this kind.

CHAPTER XXXVII

SNARES AND PITFALLS TO AVOID

THE cautious student of psychics, who desires to progress along the right lines scientifically and mathematically, must be on his guard against all possible sources of delusion and error, which may creep into his development, so that he may never mistake the false for the true, or spurious phenomena for the genuine. A few sources of error and some of the mistakes which the psychic student is apt to make will be pointed out in this chapter,—together with the means and methods of guarding against them.

First of all, do not be too credulous of the phenomena you receive and accept. If you have a chill or a nervous twitch, do not assume that this is some message or a touch from a spirit hand. It *may* be so, but you must receive good *proof* of the fact before accepting it. Should you be too credulous and accept all such incidents as genuine phenomena, you will soon be led away so far that you will become unbalanced, in your point of view.

THE OVER-NEGATIVE CONDITION

In your development do not be too negative; hold the mind always centred and conscious, as I have said, and keep the centre of yourself always active. It is only safe to abandon this in very advanced studies. Do not be too negative in your daily life or accept the advice which spirits or mediums give you to the exclusion of all else. You should reason in such matters

313

thus: "An intelligence has offered me certain advice. If that person were yet alive and offered me the same advice, would I take it?" You should accept the advice of spirits as you would that of human beings, who are merely spirits still in the flesh. In other words, as so often pointed out before, in previous chapters,—use your own judgment and discrimination on all messages received. If the messages are of an erratic nature, such as those which ask you to give up your position, go on a long journey, etc., you should be most cautious and only accept such advice after you have fully proved to your own satisfaction that it is wise and beneficial.

ABUSE OF THE "SIXTH SENSE"

Do not depend upon your "sixth sense" until you have exhausted the senses you already possess. If you refuse to let these work, you can hardly suppose that help and assistance will come from outside. No, seed will not grow in a soil that is not prepared, neither will spiritual help be planted in your mental soil if you have not worked to prepare it for this spiritual influx. As a rule, our own individual spirit is the best guide. We must consult this first. After that, if you seek additional advice and help, this may often be obtained from wise and experienced psychics, but I cannot too strongly warn the student against accepting the advice of poorly-developed mediums, either professional or amateur.

ON CHANGING MEDIUMS AND CIRCLES

It is not a good thing to change developing mediums, if this can be avoided. If you have found one medium who can assist you to develop and who is apparently doing so helpfully and rightly, stick to him through thick and thin, until his advice or help fails you. The mixture of magnetism which is introduced with change of developing mediums may be, at times, very harmful.

The same thing may be said of "circles." Once a circle of sitters is formed, this same group should sit night after night, and it is not at all a good practice to allow strangers constantly to intrude into the circle and take the places of others. If changes must be made, let one at a time assume the place of the absent sitter and let him get thoroughly familiar with the surroundings and conditions before a second change is made.

You would be wise to mistrust names of important historical persons, if they appear in your own speech or writing, or if they are obtained at séances. Our natural vanity may lead us to hope and expect that such personages may be present, but there is evidence that, in many cases, lying spirits have taken the places of those whose names they gave. In this connection it may be said that historical personages are not, as a rule, most desirable. The best help and the greatest teachings have been obtained from simple people who are now on the other side.

SENSITIVITY AND MEDIUMSHIP

Do not try from the first to develop as a medium. Try rather to cultivate your own psychic powers and strengthen your own inner nature. After you have developed psychically and spiritually in this way, you will be far better enabled to receive and transmit genuine mediumistic messages,—better enabled also to interpret them, better able to withstand the strain of mediumship, and run far less danger of obsession and other unpleasant symptoms which badly developed mediums are liable to encounter. Cultivate your psychic self, therefore, and after this has been duly trained, begin to train your mediumistic powers. Be on the lookout for evil and lying spirits, who will constantly deceive you, if you are not prepared for them, and remain too open and receptive. Study your own phenomena and endeavour to disengage genuine psychic and mediumistic manifestations from those due to your own subconscious mind. This is an excellent and very helpful practice which will prove useful to you as you progress.

Do not assume that all figures which you see are spirits. They may be thought forms, doubles, etheric bodies, or imaginary creations of your own.

THINGS A PSYCHIC SHOULD AVOID

You can only learn to disentangle this wonderful chain and separate the true from the false after months and perhaps years of study, observation and experiment. Above all, remember that symbolic figures and representations must be interpreted symbolically and should never be accepted as representing the "truth," as it actually exists. One of the great dangers to the amateur medium, as before explained, is that of extending his symbolic, intuitive impressions beyond the proper point. If he stated only what was given him, he would usually be right, but if he endeavours to interpret them himself, find their explanations, etc., he very often goes wrong. Do not "hang on," too long, so to say, to the impressions and images you perceive. Let them float before you in space, seeing and analysing them as they pass, but do not endeavour to hold them to you by the power of *your* mind. If you do so they will not only vanish and disappear, but you will be unable to retain the impression you receive and, quite possibly, the power of perceiving these images, which you now possess, will become less and less and gradually leave you. Always remember that psychic phenomena of this character cannot be *commanded*. They can only be sought and welcomed when they appear. In other words, they are "spontaneous" and not "experimental" phenomena.

HOW TO DISTINGUISH THE TRUE FROM THE FALSE

If you constantly make use of your own judgment and critical faculty in studying the phenomena which you develop or those which you may observe in others, you will build up within yourself two things: One of these is the power of judging, that is the ability to perceive the true from the false, and which, above all else, is what you, as a psychic, desire. It is difficult to explain the difference in words, but as nearly as possible it may be said that those phenomena which are innately true carry with them a sense of conviction, a feeling of warmth and familiarity, and we feel them as part of ourselves. The other phenomena, although occurring in our own minds, will seem to us cold, strange and extraneous, and when once this power to distinguish between the two types of phenomena has been developed, you have taken one of the most important forward steps that is possible for any psychic to take. Many mediums, indeed, never reach this state. Their mediumship is chaotic. It has never been developed on rational, progressive lines. But if you have done so, you may rest assured that you are not only a genuine and true medium, but that you have passed through the early stages and danger-zones which so often beset the student in the early stages of development.

HOW TO GUARD AGAINST OUTSIDE INFLUENCES

The second important step which the student takes after he has once passed this stage is that while he will be sensitive and receptive to telepathic, clairvoyant and other forms of perception and also to spirits, both in and out of the body, he will be practically impervious to harmful or malicious thoughts and influences which may be impelled against him not only on this sphere, but by the spirit world as well. If a trance-clairvoyant, during a state of ecstasy, leaves his body and wanders off into space, without having previously gained sufficient knowledge, and hence control of the situation, he is liable to be blown hither and thither (figuratively speaking) like a soap-bubble by the breezes, and will be open to impressions from all sources. These he may not feel or know at the time, but he may carry these back with him into his body and afterwards they may affect him to the detriment of his own mental and spiritual health. In other words, he has not learned to protect himself while severed from the body as he can while in it. This is one of the greatest dangers which the advanced psychic is liable to encounter, and, at the same time, after he has once learned the secret of protecting himself in this manner, he may be assured that thenceforward his progress will be most marked and rapid, not only in psychic and mediumistic development, but in the spirit-world, after he has entered it permanently, at death.

THE VALUE OF PSYCHIC DEVELOPMENT TO THE INDIVIDUAL

Psychic development is, therefore, of inestimable worth, if rightly cultivated, for the rapid progression of the individual human spirit,—just as much as the same power badly employed is harmful to the human spirit, both here and hereafter. It all depends on the manner in which these forces and powers have been cultivated and are utilized; and while too much cannot be said against their improper use, a great deal may be said in favour of their proper application and development in the right direction. It is my hope that every reader of this book will develop himself along the right lines, and that he may receive help, advice and encouragement at all stages of his spiritual unfoldment,—both here and hereafter!

CHAPTER XXXVIII

PHYSICAL PHENOMENA

THE physical phenomena of spiritualism, as distinct from the mental or psychical phenomena, are those which relate to the physical world, and in which some mechanical or physical movement of matter takes place. In clairvoyance, for example, no such physical phenomena occur, so far as we can see; but if a table be lifted into the air by supernormal means, we here come into contact with mechanical and physical forces and with these we have to reckon.

PHENOMENA WITH PHYSICAL CONTACT

We must begin at the beginning in treating of physical phenomena, and go back, first of all, to those which involve some form of *contact*. Doubtless you have seen performances of so-called mind-readers who found lost articles which were hidden in various parts of the room or hall, when one who knew their hiding-place held the psychic's hand or placed it to his forehead, etc. In most of these performances it is not mind-reading at all, strictly speaking, which we see, but what is technically known as "muscle reading,"—that is, the faint, unconscious twitchings of the muscles of the person holding the psychic's hand are felt and interpreted by him, consciously or unconsciously, and these guide him to the spot where the article is hidden. Incredible as it may appear, this is the correct explanation of these cases, and you may easily test it for yourself by asking a group of your friends to hide some object while you are out of the room, and then, when you enter, to give you one of their hands. If now you concentrate on the faint pullings and pushings which they will give you, you will be enabled to find the article in nine cases out of ten. Of course, this, like everything else, improves with practice, and you must not expect to be an expert on the first trial. Some performers, who have had years of experience, grow so proficient in this, however, that they are enabled to open safes, whose combinations they do not know, while merely holding the hand of one who does,—or even drive a cab along the streets of a crowded city, while blindfolded and holding the hand of one who can see the vehicles on the street.

THE DEVELOPMENT OF INDEPENDENT FORCE AND POWER

The next step is in Planchette-writing, where the hand, as before explained, moves at first as the result of unconscious muscular action. After a time, however, some psychic force seems to be developed and the board often continues to move about, even after the hands of the operator are removed from it.

Beyond this again we have those cases of so-called "dowsing," where the forked hazel-twig bends to and fro in the hands of the water-finder when he walks over water and metals. The simple movements which are felt at first are probably due to muscular twitchings, but as the force develops it seems to become more independent and the twig is bent in spite of the efforts to hold it.

TABLE-TIPPING AND "LEVITATIONS"

The next class of physical phenomena are those with the table. A group sits around an ordinary table, and

can tilt and tip it, as many of you have doubtless seen.
The first simple movements, here as formerly, are probably due to the unconscious muscular pressures of those
having their hands on the table; but later on, as the
psychic force develops and charges the table, it seems
to assume an independent character and the table often
continues to move when all hands are withdrawn from
it. In fact, as an expert psychic student has pointed
out "in many instances and especially under unfavourable conditions the phenomena do not rise above the
initial stage (of simple non-intelligent movements),
leaving the impression on the minds of the investigators
that the force exhibited is, if at present unknown and
unaccounted for, nevertheless a natural and a mechanical one, and that the action of independent intelligence
in connection with it, cannot be conceived. This has
been the experience and has been the verdict of even
scientific inquirers, who have not hesitated to give that
verdict to the world."

HOW THE POWER INCREASES

Such a conclusion is based upon inaccurate knowledge
and upon imperfect and superficial observation. All
experienced psychic students are aware that it is often
only after repeated and prolonged sittings, that the
full development of the psychic force is obtained and
that independent intelligence is exhibited in connection
with it and that in by far the larger number of instances
that stage of the experiment is never reached at all.
That it is, however, the ultimate issue of the experiment
is now admitted by all patient and painstaking students,
who have devoted sufficient time to the observation of
the phenomena and who have carried on their investi-

gations with an open mind and in a systematic manner. As will be seen later on, it is fully admitted that the mysterious force, thus called into operation in some unknown way, issues from the physical organism of the sensitive and the sitters, and is in itself an unintelligent force; but it is with equal confidence asserted that when it is available in sufficient quantity and is wholly detached from the physical organism, it can be and beyond all doubt *is* frequently manipulated by intelligences, independent of and other than that of the psychic and the investigators assisting in the experiment.

HOW PHYSICAL PHENOMENA ARE PRODUCED

The principle upon which many physical phenomena are based, then, is simply this: There is a vital or a nervous force existing in many of us, as described in an earlier chapter, which is usually limited to the surface of our own bodies, so that unless we touch the object in question, we cannot move it. Under certain conditions, however, this vital energy or fluid is capable of being projected outward beyond the normal bodily limits into space, and, when powerful enough, is capable of moving physical objects with which it comes into contact; or if it be a rapid outward projection of this force, it produces percussive sounds or raps well known to Spiritualists. This psychic force is often uncontrolled and then objects are moved without the knowledge of and even against the wish of the medium. We then have the so-called spontaneous "Poltergeist" Phenomena, etc. At other times this force may be guided and manipulated by the conscious or unconscious mind of the medium. Beyond this stage, again, is one in which the medium is unconscious of what is occurring,

—having passed into trance, etc., and it is then that many of the most striking physical phenomena occur. At such times complicated and intelligent physical manifestations are produced which are not due either to the mind of the medium or to any person present.

EXTERNALIZED VITAL ENERGY

We here enter the realm of genuine physical phenomena produced by spirit-intelligences. Most of the communications are obtained through raps, following a code. Playing upon musical instruments, etc., are due to this source. In other words, after a certain point has been reached, the externalized vital energy or psychic power of the medium is manipulated by an external intelligence, and they can even create forms or phantoms by utilizing it, as will be explained in the chapter devoted to Materialization.

CONTROLLING THE PHENOMENA

Very interesting experiments have been conducted in the past in *controlling* these physical phenomena, but not much success has yet been attained in this direction. There is here a wide field for experiment which the thoughtful student might enter. Thus, on one occasion, a medium who had the power of producing raps was hypnotized, and it was suggested that raps should be produced at will according to the suggestion of the hypnotist. This was completely successful. It was also suggested that raps be obtained on any article of furniture which the hypnotist would suggest. This also succeeded. The range and variety of physical phenomena are very great, including manifestations such as: raps, table-levitations, movements of objects without contact, playing upon musical instruments without apparent cause, spirit and thought-photography, materialization, slate-writing, trumpet-phenomena, etc.

THE EFFECT OF LIGHT

All physical phenomena seem to be hindered very largely by light,—either daylight or artificial light, and they can very rarely be produced except in darkness. Should you attempt to obtain phenomena of this character, therefore, it would be well for you to sit in the darkness, especially at first, and then request that more and more light be permitted as your power increases and the phenomena appear. Most mediums begin their development by seating themselves in a cabinet in a darkened room, and often it is necessary to sit in this way every evening for several weeks or even months before any phenomena appear. If you are naturally psychic, however, and physical phenomena are going to be manifested through your mediumship, you would doubtless only have to sit for a fraction of this time in order for the first manifestations to make themselves felt, and probably afterwards you would be so interested in the process that you would not count the time you spent in your development.

FIRST SYMPTOMS AND PHENOMENA

It is probable that the first indications of phenomena of this character you will receive are tiny spots of light which form before you in space, and either suddenly appear or remain stationary for some time, and then join themselves together, forming one larger light. As time progresses you will see that this light, cloudy mass will become more and more definite in outline and shape, and will probably begin to assume the shape of a phantom or form standing before you. When this stage has been reached you should concentrate your receptive faculties and endeavour to get *en rapport* with this form (for such it now is) and after a time you will doubtless be able to establish more or less intelligent mental communication and exchange messages. This will usually appear before physical phenomena become manifest, though in certain cases it may be later on. Dr. Baraduc of Paris succeeded, on several occasions, in photographing those groups of light or masses of matter which thus floated before him, and the student who has once succeeded in receiving manifestations of a like nature, might well conduct similar experiments, if he be sufficiently alert and able to do so. If not, a friend who is with him and has attended his process of development might endeavour to take these photographs at the moment when the psychic states they are vividly present before him.

There are thus two ways of cultivating physical mediumship. One is to sit in the dark; the other is to experiment more or less consciously in light or semi-darkness, and when a certain amount of power has been gained in this direction, to endeavour to transfer or carry this over into the dark séance and to transmit this power to a spirit who will thenceforth utilize it and by its aid produce physical phenomena.

DEVELOPING IN THE DARK

If you sit for physical development in the dark, you are never sure what kind of phenomena you are to obtain. In a séance this is beneficial, since you should never aim to get one type of phenomena, as before explained, for if you do you shut out by your attitude all other phenomena which might spontaneously develop. At the same time it is always satisfactory for the beginner to be able to control his phenomena a little, especially at first, and for this reason the second method of experimentation is advisable, and if desired might be carried out at the same time as the other method of development, so that the two progress side by side. If you sit in the dark you should by all means provide yourself with a cabinet, since this will tend to concentrate the force, and much less energy will have to be expended by you for the production of any phenomena you may obtain. Also you should abstain from using your will or thinking consciously of practical, every day affairs. Make the mind a blank, holding only the thought of Self, and await results.

HOW TO DEVELOP IN THE LIGHT

In developing your power for the production of physical phenomena along the other line mentioned, it is best to begin with simple experiments and gradually work up to the more complicated ones. For example: Begin with a planchette or ouija-board, placing the tips of the fingers on the board, and after it has begun to move rapidly to and fro or round and round, very gradually withdraw the hand, and see whether or not the board continues to move about. Again, when the table has begun to tip and rise into the air (two or three legs) as the result of placing your hands upon it, gradually withdraw your fingers and see whether the table remains suspended, or when it is at its highest point and you feel that it is thoroughly charged with your "fluid," drop the whole force of your being into your Will and see if you cannot levitate the table completely from the floor. Again, if raps are coming on the table upon which your hands rest, see if these cannot be obtained when your hands are removed a fraction of an inch from its surface, and if they are, endeavour to produce raps by making a motion towards the table as though hitting it, stopping short a quarter or half an inch above its surface. If you are successful, a rap or a sound in the table-top will come, following this movement.

INSTRUMENTS FOR TESTING YOUR POWER

A number of simple devices have been constructed with the object of testing mediumship in its early stages, and one or two of these you could make at home, and this would prove very helpful to you. Thus: You might suspend a small pith or cork-ball by means of a silk thread, five or six inches long, from a hook. If now you place the fingers of one hand almost touching this ball and leave it there for some moments, you may, if successful, succeed in causing this ball to move either towards or away from your fingers as you will. This is a very useful little experiment which may be tried on many occasions and will be found very beneficial in developing simple physical phenomena. Another device which may be employed is the following: Procure a straw such as used at the soda-fountains, and pass a needle through it directly in the centre. Press the lower end of the needle into a large, flat cork; see that the straw revolves easily upon the slightest pressure. Place your fingers nearly touching one end of the straw and *will* that it shall move either to the right or to the left. This instrument has proved very successful in many cases and will probably prove more sensitive than the last.

There are more complicated scientific instruments which have been devised to test the externalization of the human fluid, and the power of the will. These instruments have been used with great benefit by many scientific students.

HOW TO BEGIN

When the student has progressed thus far, he is ready to try his first experiment in the movement of physical objects lying on his table. Begin with a very small, light object, such as a cork. Do not choose any metal object. Place the finger-tips of both hands on either side of the object, nearly touching it. Wait until you feel distinct tingling sensations in the fingers, and if this sensation extends to the elbows, or even to the shoulders, so much the better. Endeavour to construct by your will and in imagination, so to speak, a fine thread or hair, composed of psychic rays, passing between your fingers and supporting the object in question. Concentrate on this for some moments before you make any physical movement. Then, very slowly raise the fingers and see whether the cork is influenced to follow the upper direction of your fingers. If so, you have begun your course of physical mediumship! As this initial experiment is very important, it would be well to dwell upon it at somewhat greater length, since nothing is so discouraging to the beginner as innumerable *tests* and *experiments* of this kind which fail one after the other. (Of course, perfectly non-mediumistic persons will continue to fail, but natural psychics will not.)

HOW TO OBTAIN THE FIRST PHENOMENA

We have seen in an earlier chapter that the aura extends from the body and particularly the finger-tips, and that this human fluid is capable of projection at will. Now, it is this fluid which is the basis or substance out of which the psychic threads or rays are spun, and these threads, when they have stretched from finger to finger, and gained sufficient solidity, are capable of lifting quite heavy objects. Dr. Ochorovitz, who has studied these rays for years, calls them "rigid rays" and asserts that his medium, Mademoiselle Tomczyk, can by an effort of will construct a psychic thread so strong that it can be heard scraping against solid objects and even *seen* occasionally,—yet it does not exist as a physical reality, for the space between the fingers and the object may be cut without severing the connection!

Now, these psychic threads are woven not of physical but of etheric or astral matter, and as we do not know as yet how to mould or manipulate this accurately, we have to do the best we can by the power of the human will. The process to be followed therefore is: first, vivid imaginary construction of these rays or threads; second, projection of the vital fluid; and third, the weaving of this together into the rigid rays by an effort of will. If the student can follow this process and persistently carry out the instructions, he will doubtless succeed in time in moving small, light objects,—that is, if he is at all gifted with this phase of mediumship.

HOW TO CONSTRUCT THE VITAL THREADS OR "RIGID RAYS"

The details of this process may now be given. First of all, place yourself in a relaxed, restful condition.

Then think intently of the threads or rays which you wish to produce. Imagine these just like any other threads coming from your finger-tips and becoming more and more dense and solid as they emerge. Think of the strips of fluid you saw between your finger-tips in trying the experiments mentioned in Chapter XXV, devoted to "the human fluid." When you have formed these vital rays clearly in your mind and have them all ready to project, so to say, extend the fingers and by a strong effort of will, endeavour to project this energy, into space, beyond the finger-tips. After a very few trials, you will doubtless begin to do so. This you will feel in the form of "pins and needles" sensations in the finger-tips. They will also get warm, perhaps perspire. When this second stage has been reached, you are ready to proceed with the third. The fluid thus projected is not in the form of rigid rays or threads, but rather a vaporous mass, a soft cloud,—if the term be allowed,—and you must toughen and strengthen this by will-power. After the projection has taken place, think and *will* intently that this shall happen, and is happening, and at the same time imagine your consciousness in your finger-tips themselves, moulding and "toughening" these vital rays. If you do this, you will surely succeed in time,—provided you go at the exercise in the right manner, and "stick to it" persistently.

TRANSFERRING THE POWER

When the student has progressed thus far, the final step must be taken, namely the transferring of this power to the control of a "spirit" or outside intelligence. This is a very delicate and subtle process, which is very little understood, even by mediums. The best process

is gradually to develop the power of going into trance
coincidentally with the development of these physical
phenomena. Once you have gained the power of projecting your fluid at will and moving material objects by
its aid (which is probably attained by an extreme effort
of will) you should endeavour to hand over this manipu-
lative power to another intelligence. You cannot do this
consciously so you can only hope that the transference
will take place when you have passed into trance. You
should endeavour, therefore, to pass into trance while
actually conducting the above mentioned experiments,
and the proof of the fact that this transference does take
place is found in the fact that the most striking physical
phenomena at a séance always occur when the medium *is*
in deep trance. The deeper the trance, the better the
phenomena! In other words, the more the medium's
will is in abeyance, the more opportunity is there given
to the external will of the spirit to become active and
bring about the required results. This fact is very
strikingly proved by nearly all the best physical mediums
in the history of Spiritualism.

GATHERING VITAL ENERGY FROM THE CIRCLE

If you are unable to move material objects alone, you
may perhaps be enabled to do so after gathering strength
from others. You may do this either by forming a chain
and gathering this energy by an effort of will, before
you make your experiment, or by placing your hands in
position and asking the two members of the chain nearest to you to place their hands upon your temples, or
one on your forehead, and the other over the solar
plexus. In this way a vital magnetic current is estab-

lished which may greatly add to your powers and enable you to move objects and produce phenomena where you would otherwise fail.

CHAPTER XXXIX

SPIRIT AND THOUGHT-PHOTOGRAPHY

SPIRIT-PHOTOGRAPHS are based on the belief that there is a spiritual body, resembling in appearance the physical body, which is sufficiently solid to be photographed by means of the camera and sensitive plates. Usually more than this is necessary,—namely the presence of a medium or psychic, possessing the peculiar power of rendering the spiritual body apparent to the camera. The medium seems to act as a sort of connecting link or intermediary between the body and the photographic plate, though the exact nature of the mediumistic influence is as yet unknown. Here is a field for study by expert photographers and by scientists, to ascertain its limits and extent.

HOW SPIRIT-PHOTOGRAPHY IS POSSIBLE

To many it may appear incredible that any spiritual body is sufficiently material to be photographed by the camera, for it would mean that this body is capable of reflecting light-waves, this being the primary necessity in obtaining photographs at all. Yet, as Sir Oliver Lodge has pointed out, there is hardly anything more incredible in this than in taking the photograph of the reflection of an object in the mirror, which is quite possible. In this case there is no solid object photographed —merely the reflected light-waves which are themselves intangible and invisible.

We know from experiment that the photographic

camera is far more sensitive than the human eye. Physicians tell us that it is possible to photograph an eruption on the body before it actually occurs, that is, before it is visible to us (such as smallpox).

On the other hand, it is also possible to photograph thousands of stars in the heavens, which are invisible to the eye, even with the most powerful telescope. A photographic plate can therefore detect objects insensible to the eye, and hence it is reasonable to suppose—inasmuch as spiritual bodies doubtless exist, but are just beyond the range of our vision,—that the camera should be quite able to detect them, and spirit photographs are the result.

TWO SOURCES OF ERROR, AND HOW TO GUARD AGAINST THEM

In obtaining spirit-photographs you must be on your guard against two possible sources of error. The first is, that you are liable to see faces and likenesses in the photograph which do not really exist at all—you construct them in imagination as you would faces in a coal-fire. The second danger to be avoided (if you are dealing with a professional spirit-photographer) is that of fraud. There has doubtless been much trickery in this department in the past, and if you wish to be sure that you are not victimized you should take your own plates with you, see them inserted in the camera and watch their development after the picture has been taken. Even in this case you are liable to be imposed upon, unless you are very careful.

HOW TO BEGIN YOUR DEVELOPMENT

The most satisfactory course to pursue is to experiment yourself and not depend upon a professional spirit-

photographer for your results. If you are at all sensitive and persevering, you will doubtless obtain genuine
spirit-photographs at the end of a certain period of time.
Many hundreds of persons have done so and there is
no reason why you should not, if you are determined
to obtain them.

The best method is to sit privately with a friend of
yours, who is both sympathetic and more or less mediumistic, and hold a short séance, seated at the table, before
you begin your experiments in photography. If you obtain messages by means of tippings of the table, raps,
automatic-writing, etc., so much the better, and if intelligent communication is thus established, ask your spirit-
friends to appear for you on the plate when the experiments are being held. They may promise to do so, but
fail to appear. Do not be discouraged by this, as they
may be perfectly willing to help you, but for some reason or other are unable to make their forms visible on
the photographic plate. If you persist, however, you
will doubtless obtain interesting results in a short time.

HOW TO TAKE THE PHOTOGRAPHS

After this preliminary séance, you should seat your
subject in a chair against a dark background, and focus
the camera as you would were you taking his picture
in the ordinary way. The photographic plate should, if
possible, be held by both of you between your hands in
the dark room, before being inserted in the camera, so
as to get it impregnated with your "magnetism." After
he has taken up his position, and the camera is properly
focussed, you should then ask your spirit-friends to come
and appear on the plate, if possible. Do not exercise
your will, however, nor think of any special object in

particular, nor any person, but make your minds negative. If positive, you are quite likely to obtain thought-
photographs instead. Ask your invisible helpers to give
you some sign, if possible, such as three raps when they
are ready to appear, etc. If you obtain these, take the
picture at once, if not, sit until you get into the requisite
mental condition, then take the photograph and afterward develop it carefully. It is improbable that you
will obtain any definite results for the first few experiments,—but many do, even from the start, and this is
doubtless one of the most promising of all the fields of
psychic investigation for the student to enter.

"RADIOGRAPHS," AND HOW TO OBTAIN THEM

The next thing to do is to endeavour to secure photographs of the rays or aura of the human body. These
impressions on the photographic plate are secured comparatively rarely, for the reason that the body of the
subject must become "radio-active" to some extent before an impression of this kind is possible. Such pictures are consequently called "Radiographs," and a
number of these have been obtained by Dr. Ochorovicz
of Poland. The rays in question, which impress the
photographic plates in such cases, seem to emanate from
the etheric double and not from the physical body, for
the reason that they do not follow the anatomical dis-
tribution of the nerves of the body. The "double," detached after the manner described in Chapter XXVI,
can often affect the plates in this way, and spirits can do
so, but it is not common for the human body to be able
thus to affect them.

HOW TO OBTAIN THOUGHT-PHOTOGRAPHS

The third and most interesting phase, in a sense, for the experimenter is that of "Thought-Photography." The most sensitive plates that can be procured should be used for this purpose and the experiments conducted in the dark, (as indeed should the Radiograph experiments). The plate may be held between the palms of the hands or placed against the forehead or over the solar plexus, next to the skin, and must be left there for a considerable time—half an hour or longer, if possible. During this time the subject should think intensely of a certain figure or object, such as a cat, a chair, a ship,—as the case may be. He should keep this before his mind vividly and intensely and never allow it to become blurred or indistinct. Holding it there by an effort of will, he should next endeavour to impress this upon the photographic plate, and should also try to feel inwardly the process going on within him—the flow of the magnetic current to the spot beneath the plate, etc.

ANOTHER WAY TO PRODUCE THOUGHT-PHOTOGRAPHS

Another way of obtaining thought-photographs is to place a plate wrapped in black paper, or placed in an opaque black envelope, on the table, and over it place the finger-tips for some time,—usually from 5 to 10 minutes. Then think or will that a certain thought or image will be impressed upon the plate; and if you are at all developed along this line, the impress will be left on the plate, through the paper. Any object can be selected—a round ring of light, a triangle, a face, etc. It is best to begin with simple objects, because the mind seems able to impress this upon the plate more readily and clearly than a more complex object, of which it cannot form so clear an outline.

You must not be disappointed if you do not succeed at first in this, and you may have to develop (and thus spoil) a number of plates before you get any impression at all upon them. The first thing you will get, probably, will be a spot of light, or a series of small spots, as the *fluid* finds its way through the opaque paper, unto the plate. You must remember that the human fluid is the instrument or intermediary, through which photographs of this character are made, and hence you must learn the art of the projection of this fluid, as outlined in the chapter devoted to "Physical Phenomena," before you can hope successfully to impress a photographic plate. Once you have done so, the rest will be simply a matter of development; and you will find it one of the most interesting and fascinating subjects for investigation in the whole realm of psychics.

PHOTOGRAPHS OF PSYCHIC FORMS AND EMOTIONS

In many cases photographs of emotions have been successfully taken, especially of late, and Monsieur Darget has narrated a number of experiments of this character to the French Academy of Sciences, which has accepted his report as authentic. It is thus evident that Thought-Photography has at length claimed a place in the scientific world, and, this being so, it is only a matter of careful experimenting on the student's part before he obtains photographs of this character.

An interesting series of experiments might be tried by the scientifically minded inquirer, namely, to obtain photographs of mediums in trance, while they

are obtaining Automatic-Writing, Crystal-Gazing, etc.,
and also of those who are on the point of dying. Such
experiments would doubtless reveal many changes in the
Aura, and also the presence of Thought-Images and possibly Spirit-Forms which would otherwise be quite unsus-
pected by those present.

CHAPTER XL

MATERIALIZATION

MATERIALIZATION means the process of rendering solid or material, for a longer or shorter time, bodies through which disembodied spirits may function and communicate. Materialization usually occurs at séances in which a group of people are gathered together, and rarely or never when the medium is alone. The reason for this is probably that the necessary conditions are lacking, these being chiefly the lack of sufficient vital energy, which is drawn from the circle by the medium and utilized for the purposes of materialization.

THE MARVELS OF MATERIALIZATION

Many factors play a part in this mysterious phenomenon. Considered from the physical or material point-of-view, there is the reality of the phantom, and from the psychological or mental point-of-view, there is the mind of the materialized entity to account for. If we were always sure that the materialized figure were really the person it claimed to be, this latter difficulty would be overcome, but as we shall see later, there are many objections to this simple view of the case in all instances, and thus the problem is rendered more complex.

From the purely physical point-of-view, the phenomena of materialization are the most baffling and the most mysterious in the whole realm of Spiritualism. A

few minutes before, nothing existed in the cabinet, save
the entranced medium. Now, there is a solid, tangible
form possessing all the properties and appearances of
matter, often having solid flesh and bones just as a
human being would,—the flesh being warm and life-
like, the hand possessing nails, hair, etc., like an ordinary
hand, and being apparently composed of cells and tissues, such as any material body would be composed of!
How account for this? It is surely one of the most bewildering and incredible facts in Nature.

THE NECESSARY FACTORS TO INSURE SUCCESS

From the point-of-view of spiritualism and psychic
development, many factors play a part. There is first
of all the physical body of the medium, secondly his vital
magnetism, thirdly the magnetism of the sitters forming the circle, fourthly, magnetism from disembodied
spirits, which mingle together and help to create the
phantoms that appear at séances. The vital energy
which seems to be drawn from the circle, and chiefly
from the medium, during the séances, is utilized or
manipulated by the disembodied spirits, who build up
by its aid the materialized form we see before us. This
is a very difficult and complicated process and not all
spirits are competent to do this. For this purpose what
are known as ''spirit-chemists'' are often employed, those
who possess the knowledge of how to build up these
forms. In the deepest stages of trance, when the medium
is unconscious, the communication through materialized
figures becomes clearer and clearer and the forms more
dense and material. This is true of many psychic phenomena: the deeper the trance, the better the results obtained.

ETHERIALIZATION AND TRANSFIGURATION

In the lighter stages of trance, however, only portions of the figure may develop, such as hands, faces, etc., or very shadowy and vaguely defined outlines of human forms. These latter are not, strictly speaking, materialized but are known as "etherialized" forms. They are less solid than the materialized figures, and it is often possible to pass the hand and arm through one of these figures without disturbing it. In the case of the materialized figure, on the other hand, they are just as solid and tangible as any human form and it would be impossible to make any other solid object pass through any part of them. In many cases the physical body of the medium is more or less altered by the spirits without any other phantom being created. Such cases are known as "transfiguration." When the figure created at the séance is not dense and fully formed, it does not possess either a complete or matured intelligence. It is not "all there," so to speak, mentally or physically.

HOW SOME FORMS ARE CREATED

There is evidence to show that many of these forms are created by the will of the medium or by discarnate spirits, and that they are more truly thought-forms than materialized spirits. Again, many of these figures are "doubles" or "astral bodies" belonging to living people, who happen to appear at the séance, or projections from discarnate spirits. In such case, the intelligence manipulating the phantom is not that of a mature spirit but is a creation, so to speak, elaborated by the subconscious thoughts of the medium or by the mentality of the sitters, forming the circle. The psychic atmosphere created by the minds in the circle has, in other words, produced the mind of the phantom in the same way that the combined vital magnetism of the sitters has produced the material body of the apparition.

HOW MATERIALIZATION IS ACCOMPLISHED

The process of materialization seems to be somewhat as follows:—The vital energy being drawn from the sitters into the body of the medium, the latter projects it outward into space, together with a large portion of his own vital energy, or it is drawn out by the operating intelligences. When in space, at a short distance from the medium's body, this vital energy is moulded, so to say, into the shape of the materialized form. It is built up or created by the operating intelligences. Between this form and the medium's physical body there exists a subtle connection or "rapport" which has been described as a thread or bond of union, though it is not a physical connection of any kind or one that has ever been detected. Yet, that such a connection exists is proved by the phenomena of "repercussion," referred to in Chapter XXXVI, where it was shown that any injury done to the projected form reacted upon the body of the medium and left its mark upon it, just as though the physical form had suffered the injury. This is one of the most striking phenomena in the whole realm of spiritualism, and a case of this character is thus vividly described by the Ven. Archdeacon Colley in his address on "Spiritualism" before the Church Congress which met in October, 1905, and subsequently published by him in pamphlet form. He then said:—

"REPERCUSSION"

"He (the materialized phantom) seemed to be interested in everything around him, walked up and down the room, taking up various articles to examine them, as would be natural to one of ancient race now in the midst of modern environment. Presently he espied and brought from the side-board a dish of baked apples and I got him to eat some. Our medium was at this time six or seven feet away from the spirit-form and had not chosen to take any of the fruit, asserting that he could taste the apple the Egyptian was eating. Wondering how this could be I with my right hand gave our abnormal (?) friend another apple to eat, holding a bit of white paper in my left hand outstretched toward the medium, when from his lips fell the chewed skin and core of the apple eaten by the Mahedi; here it is before me now after all these years in this screwed-up bit of paper for any scientist to analyse."

In this instance the phenomena of repercussion was very interestingly demonstrated. The method of the materialization of the figure was thus described by Archdeacon Colley in his lecture:—

HOW THE FIGURES ARE FORMED

"When, in expectation of a materialization, . . . there was seen steaming as from a kettle-spout through the texture and substance of the medium's black coat a little below the left breast toward the side, a vaporous filament which was almost invisible until within an inch or two of our friend's body. Then it grew in density to a cloudy something. There would then step forth timidly a figure—as did this little maiden. . . . She was

naturally a companion for others of our frequent psychic
visitors. For 'as a cloud received one out of their sight'
when the disciples at Bethany gazed on their ascending
Lord, so, as from a cloud thus inexplicably evolved from
the medium, came our materialized friends, and vanished
again to invisibility in a cloud (sucked back within his
own body) when they were withdrawn from us, wist-
fully gazing on the mysterious departure and noting
this or that particular phase of it within a few inches
of the point of their inscrutable disappearance and evan-
ishment.''

THE CLOTHES OF MATERIALIZED FIGURES

The question is often asked: ''How is it possible for
spirits to become *clothed?*—the old question of the
''clothes of ghosts'' being often raised among material-
istic sceptics of the last century. The same question
might be raised against the clothes of materialized figures. But there is a ready answer to this which fully
explains it. Those who deny and ridicule the possibility
of *materialization of raiment* (as well as bodies) might
ask themselves the question, ''Whence came the clothing
which Christ wore after his resurrection?''—for we are
distinctly told that the Master's raiment had been parted
among the Roman soldiery ''and upon his cloak had they
cast lots.'' This historical incident furnishes us with
an illustration of the case in point, and the reality of
this fact is amply borne out by many modern instances
of a like character.

HOW TO BEGIN YOUR DEVELOPMENT

In sitting for materialization, the medium should sit
inside the cabinet, which should not be too large, so as

to concentrate and focus the energy obtained from the circle. The medium should sit on a cane bottomed chair, sufficiently comfortable to afford perfect relaxation when the trance supervenes. At first the medium should hold the hands of those in the circle, but after a time these may be released. The light should be almost totally extinguished, for reasons given before in this book.

It must be remembered that there are all kinds of light, visible and invisible. We also have infra-red rays and ultra-violet rays,—the former being below the lowest form of visible vibration, and the latter above the highest. It is because red is so low in the scale of vibration that mediums employ it during the séance. Photographs may be taken by infra-red and ultra-violet light.

Light has a very disintegrating effect on these subtle forms and would doubtless serve to disintegrate many of the materialized forms upon their initial appearance. The medium should make his mind as blank as possible, holding only the central idea of self, and mentally call upon his spirit friends to help in the production of phenomena.

EARLY SIGNS AND PHENOMENA

Among the initial sensations which the medium will experience are, probably, flashes of heat and cold, blackness before the eyes (in which possibly there may be specks of light dancing hither and thither) and a "cobwebby" sensation over the hands and face, which is almost invariable and very noticeable.

Madame D'Esperance, a materializing medium of international fame, has stated that in her experience this cobwebby sensation was present on practically every occasion. Speaking of the phenomena and symptoms of the process, she says:—

"If a few persons have gathered together in a half-darkened room, the emanations from their bodies can be seen by many—not necessarily clairvoyants. It appears as a slightly luminous haze about the head, shoulders and sometimes the knees and feet. Frequently it gathers slowly at the fingers, increasing in density till it resembles a slight transparent film of slightly luminous cotton-wool. This is often perceptible to the eyes of all, but it offers no resistance to the touch. By some force of attraction, either inherent or exerted upon it by some outside agency, this mass appears to mingle and draw together, to become more dense and at this stage has been found to be decidedly perceptible to the touch. It resembles as nearly as can be described the gossamer web, seen on trees and bushes on an early summer morning."

THE SENSATION OF "COBWEBS," AND WHAT IT MEANS

"Many persons in a materialization séance are sensible of a feeling as of cobwebs being on their faces and hands. I have myself not only felt the sensation, but when brushing my face or hands have distinctly felt what seemed to be fine filaments of the gossamer which clung to my fingers. The attention of the sitters has been frequently drawn to this almost impalpable substance which has vanished as soon as the light has been brought near it. . . . This emanation from the sitters in a séance is generally, if not always, accompanied by a sensation of chill or draft, similar to that felt by a person in a slightly feverish condition. . . . The head will be hot, there will be a heavy throbbing in the temples.

The hands, feet and other parts of the body will be cold to the touch. . . . The medium by the exercise of his will can at any time prevent manifestations,—in fact the opposition of any person in a circle will act as a hindrance to the work of the unseen operators.''

WHY SOME FORMS RESEMBLE THE MEDIUM

As a rule, when full materializations are accomplished the medium is entranced so deeply that he cannot remember the process of the production of the forms. In the earlier stages of trance, the mind should be concentrated on the creation of forms of this character, but after it has reached a certain stage, you may safely turn over the process to your spirit-friends. In some instances, the medium's double becomes detached from the body and appears to those forming the circle as a materialized figure, though it is not such in reality. If such a figure be photographed or closely examined, the striking resemblance to the medium is easily seen, though it is not the medium, who may be seen entranced within the cabinet. Lack of knowledge of this fact has given rise to the false belief that in cases of this character the medium himself was consciously personating the spirit, but the true explanation is that the double has been liberated during the séance and has thus appeared to the sitters as an independent being.

The phenomena of materialization, as before said, are amongst the most interesting in the whole realm of the super-normal, and will well repay careful study and prolonged experimentation on the part of the student.

CHAPTER XLI

ADVANCED STUDIES

THE subject-matter and advice contained in the present chapter is advanced only for those who have carefully read through and practised the preceding chapters of this book. Those who have not done so are strongly advised not to undertake some of the experiments herein described unless they have carefully carried out the instructions contained in the earlier chapters, and particularly the warnings herein given. These advanced studies are suitable only for those students who have succeeded in attaining a certain mastery of the inner self, and who have developed a certain amount of psychic force or power which is under their own control. In a certain sense they may be considered more or less dangerous, but they are not so to one who has progressed sufficiently to be in a position to follow them. Progress is necessary in psychic development as in every other field of endeavour, and those who have gone thus far should try to advance their powers and faculties yet another step forward into that vast and mystic beyond which encircles us on every side—not only in the life to come but here and now.

CULTIVATING THE "SIXTH SENSE"

The first thing for the student to do is to cultivate as far as possible his "Sixth Sense," already mentioned briefly in Chapter XIX (devoted to the Cultivation of

Sensitiveness). This sixth sense is a general feeling of "awareness" of surrounding powers and entities—a knowledge which is not dependent on any of the five senses. Some of the preliminary exercises for cultivating this sense have already been given, and we shall now proceed to give a few more, leading the student yet further along the path to self-realization and power.

He should first of all begin with deep breathing-exercises, accompanied by certain psychical processes and practices. The process of taking the "complete breath" has already been described in Chapter VI, and while the student is in the relaxed condition, previously mentioned, he should concentrate his mind and carry out the following psychic formula:—

PSYCHIC BREATHING EXERCISES

Breathe rhythmically, until the rhythm is perfectly established, then, inhaling and exhaling, form the mental image of the breath being drawn up through the bones of the legs and then forced out through them; then through the bones of the arms; then the top of the skull; then through the stomach; then through the reproductive region; then as if it were travelling upward and downward along the spinal column, and then as if the breath were being inhaled and exhaled through every pore of the skin,—the whole body being filled with "prana" (vital energy or life),—and, breathing rhythmically, send the current or "prana" to the seven vital centres in turn, as follows, using the same mental picture as in the previous exercises.

First, to the very end of the spinal cord; second to the reproductive region; next to the centre of the abdomen; next to the solar plexus; then to the heart; then to the throat; then to a spot between the eyes, low down on the forehead; finally, to a spot at the very top of the brain. Finish by sweeping the current of "prana" to and fro from head to foot several times.

HOW TO AWAKEN THE "CHAKRAS" OR 7 VITAL CENTRES

These seven vital centres in the body are known as "chakras" and have very great interest and importance in all higher psychic development and in all occult practice. It is upon the awakening of these Seven Centres, in fact, that all the higher clairvoyance and psychical faculties depend. They are supposed to be the links of connection between the physical and the astral bodies, and if they are not awakened in precisely the right order, and in the right manner, grave difficulties may result; while, on the other hand, if they are awakened correctly, the student who has done so is instantly gifted with extraordinary clairvoyant and higher psychical faculties, —enabling him not only to see the past and the future, but also all those spiritual beings who are constantly around him,—the thoughts and emotions of others, pictures of their past lives, etc. In other words, much depends upon the awakening of these centres! In Eastern Philosophy they are symbolized as "Lotus Flowers," and the highest and last in the brain is called "the Thousand and One Petalled Lotus."

IMPORTANCE OF AWAKENING IN THE RIGHT ORDER

The Vital Energy which passes upward through these centres is symbolized as a Fiery Serpent which, in passing upward, animates each in turn, and wakes them into activity, and it is highly important that this current of energy should pass through each centre in the *right order*, as before said. The sensation of warmth and a faint prickling as of "pins and needles" is felt at the moment of awakening each of these centres. In Sanscrit the word "Kundalini" (literally meaning "the coiled-up") is employed. This "serpent," when fully aroused and activated, leads not only to the awakening of the higher psychical faculties before mentioned, but also to others of a still more startling character.

Swami Vivekananda in his "Lectures on Raja Yoga" (p. 91), gives the following psychical exercises which should be practised in connection with this psychical un-foldment and development:—

THE SACRED WORD "OM" AND MEDITATION

"Sit straight, and look at the tip of your nose. By controlling the two optic nerves one advances a long way towards the control of the arc of reaction, and so to the control of the will. . . . Imagine a lotus upon the top of the head, several inches up, and Virtue as its centre, the stalk as knowledge. The eight petals of the lotus are the eight powers of the Yogi. Inside the stamens and pistils are renunciation. . . . Inside of the lotus think of the Golden One, the ALMIGHTY, the Intangible, HE whose name is OM, the Inexpressible, surrounded with effulgent light. Meditate on that. Think of a space in your heart, and in the midst of that space think that a flame is burning. Think of that flame in your own Soul, and inside that flame in another space, effulgent, and that is the Soul of your Soul,—God. Meditate on that in the heart. He, who has given up all attachment, all fear, and all anger, he who has taken refuge in the Lord, whose heart has become purified, with whatsoever desire he comes to the Lord, he will grant that to him."

INTERNAL OR SPIRITUAL RESPIRATION

Another valuable practice in connection with breathing is that which is known as "internal" or "spiritual respiration." The idea is based upon the belief that, in addition to our physical lungs, there are also spiritual lungs, and that just as the physical lungs receive energy and are purified by the air we breathe,—so also are the spiritual lungs energized and filled by the power of spirit, when accompanied by suitable psychical and mental processes. The power of the word "OM," so often repeated in Eastern Philosophy, may be perceived faintly by any one pronouncing the word slowly, several times in succession, when it will be seen that it has a peculiar psychical effect upon the individual, and that it sets up remarkable rhythmic vibrations throughout the whole being, which become more and more noticeable as the word is repeated. This is the most holy word of the Vedas or sacred books of the East, and its symbolic meaning is "The Supreme Being," "The Ocean of Knowledge" or "Bliss Absolute."

SEEING WITH ANY PART OF THE BODY

One other valuable exercise which should be practised is that of seeing, or endeavouring to see, with any part of the body, as though eyes were situated at any point upon which you concentrate your forces, and that you were actually looking outward from that point. This power has been cultivated to an extraordinary extent by some of the Eastern Adepts and is recorded as happening spontaneously now-and-then, even now, in the East. The power is cultivated by an effort of attention, coupled by will, and should be preceded by the practice of travelling around the body in thought,—mentioned before in this book,—and then holding yourself consciously on one particular point, in your circuit of the body, and concentrating yourself on that point.

At this stage of your development, you may begin to practice an exercise which would be of great benefit, not only to yourself but to others also. After you have fallen asleep,—and the astral body is thereby loosened from the physical body, you should learn to make use of this astral body during the hours of sleep, and send it on journeys, to help those who may be in need of this help. You may, after a certain amount of effort, thus project the astral body, and cause it to retain full self-consciousness. When this has been acquired this projected body can assist those who have recently died, comforting and consoling them, and can carry messages from such a person to those still living. It can assist those in danger, and help along humanity in a thousand different ways. When you have learned to project your astral body in this manner during sleep, you are known as one of the "Invisible Helpers" and many persons are said to make it a business to perform at least one good action every night, during sleep.

THE DEVELOPMENT OF COSMIC CONSCIOUSNESS

Two remarkable psychical manifestations will result from these spiritual practices, if correctly and carefully performed. The first is the enlargement of the Self until it attains a vast area, so to speak, which has been called "Cosmic Consciousness," by those who have ex-

perienced it. This consciousness is a step higher than
human consciousness,—just as the human is a step higher
than the animal,—and enables us to perceive truth and
spiritual reality behind the universe, in addition to
stimulating remarkable psychic powers. Such realities
as the "fourth dimension," which are usually quite incapable of being appreciated by our finite senses, are
said to be clear and intelligible to those who possess
Cosmic Consciousness, and the connection between spirit
and matter is also clear to them.

POWER OVER ANIMATE AND INANIMATE MATTER

The second remarkable development from the awakening of these higher spiritual faculties will be the
greater power you possess over animate and inanimate
nature. You will find that you exert a peculiar influence over all animals with whom you come into contact, and that they not only know and understand you,
but, if the animals are wild, they will not harm you
in any way. It is stated that many of the Yogis of
India can walk uninjured through dense jungles filled
with tigers and venomous snakes. These facts throw a
new and interesting light upon the account of "Daniel
in the den of lions." Doubtless all the Biblical narra-
tives of this kind can be rationally accounted for, when
we have acquired sufficient knowledge of psychic and
spiritual science. Even the case of the three men who
were cast into the fiery furnace and escaped uninjured!
Several mediums have done the same thing on a small
scale. Sir William Crookes has reported that he has
seen the medium D. D. Home extract red-hot coals from
the fire and hold them in his hands without injury.
Similarly the magicians or witch-doctors of many of the

savage tribes can walk over glowing coals or red-hot
embers without being burned,—after they have under-
gone certain religious rites and preparations.

In addition to this, you will have increased power
over inorganic matter, so that you can move objects
without contact, with comparative ease, and cause phe-
nomenal changes to take place in those objects. You
will find that you have, in an almost perfect degree, the
power of "self-projection,"—that you can leave your
body and enter the astral plane at will, exploring it and
observing its denizens.

CREATION BY THE POWER OF WILL

Finally you will be able actually to *create* by the
power of your thought, forms and objects which are external and apparently objective. In other words, you
will have learned to "create" by the power of the will,
—and this is one of the greatest achievements gained by
the advanced student of the occult. Phantoms, Apparitions, Thought-Forms, etc., are created in this way.

It is impossible at this time, to enter more deeply into
these questions. Higher exercises of this kind, to be
explained fully, as they should be, would require a further Course of study; and it is my intention to follow the present work with a second one,—which will
contain more detailed advice as to the development of
the higher psychical and spiritual faculties. For the
present, I must leave the psychic student here, at the
end of his preparatory studies,—wishing him success in
his efforts, in the attainment of psychic power. If the
student will but follow the directions contained in the
present work carefully, and at the same time pay due

attention to the advice contained therein, he will be enabled to develop his psychic powers to a very great extent, and will thereby be fitted to undertake still more advanced studies, which will be taken up very fully in a subsequent work.

THE END